# BEAL'S CONJECTURE

# BEAL'S CONJECTURE

Ran Van Vo

*AuthorHouse™ LLC*
*1663 Liberty Drive*
*Bloomington, IN 47403*
*www.authorhouse.com*
*Phone: 1-800-839-8640*



*Published by AuthorHouse 03/27/2014*

*ISBN: 978-1-4918-9944-1 (sc)*
*ISBN: 978-1-4918-9943-4 (e)*

*Library of Congress Control Number: 2014906047*

# Contents

# BEAL'S CONJECTURE
# And PRIZE

# INTRODUCTION

Billionaire banker Andrew Beal has conjectured that there are no solutions to the equation

$$a^x + b^y = c^z$$

If a, b, and c are co-prime integers, and x, y, and z are all integers > 2.

*The problem is to prove the Beal Conjecture, or find a counter-example.*

Andy Beal initially offered a prize of US $5,000 in 1997, raising it to $50,000 over ten years, the prize is now this: $1,000,000 for either a proof or a counterexample of his conjecture

Andy Beal spent a lot of time to prove Fermat's Last theorem,

> *"Andy Beal and a colleague programmed 15 computers and after thousands of cumulative hours of operation had checked all variable values through 99. Many solutions were found: all had a common factor in the bases. While certainly not*

*conclusive, Andy Beal now had sufficient reason to share his discovery with the world."*

1993 Andy Beal stated as follows:

*"BEAL'S CONJECTURE: If $A^x + B^y = C^z$, where A, B, C, x, y and z are positive integers and x, y and z are all greater than 2, then A, B and C must have a common prime factor"*

Ex:

$$3^3 + 6^3 = 3^5$$
$$3^9 + 54^3 = 3^{11}$$
$$3^6 + 18^3 = 3^8$$

# *The Beal Conjecture*

## *Background*

*Mathematicians have long been intrigued by Pierre Fermat's famous assertion that $A^x + B^x = C^x$ is impossible (as stipulated) and the remark written in the margin of his book that he had a demonstration or "proof". This became known as Fermat's Last Theorem (FLT) despite the lack of a proof. Andrew Wiles proved the relationship in 1994, though everyone agrees that Fermat's proof could not possibly have been the proof discovered by Wiles. Number theorists remain divided when speculating over whether or not Fermat actually had a proof, or whether he was mistaken. This mystery remains unanswered though* *the prevailing wisdom is that Fermat was mistaken. This conclusion is based on the fact that thousands of mathematicians have cumulatively spent many millions of hours over the past 350 years searching unsuccessfully for such a proof.*

*It is easy to see that if* $A^x + B^x = C^x$ *then either A, B, and C are co-prime or, if not co-prime that any common factor could be divided out of each term until the equation existed with co-prime bases. (Co-prime is synonymous with pairwise relatively prime and means that in a given set of numbers, no two of the numbers share a common factor.)*

*You could then restate FLT by saying that* $A^x + B^x = C^x$ *is impossible with co-prime bases. (Yes, it is also impossible without co-prime bases, but non co-prime bases can only exist as a consequence of co-prime bases.)*

## Beyond Fermat's Last Theorem

*No one suspected that* $\boldsymbol{A^x + B^y = C^z}$ *(note unique exponents) might also be impossible with co-prime bases until a remarkable discovery in 1993 by a Dallas, Texas number theory enthusiast by the name of D. Andrew "Andy" Beal. Andy Beal was working on FLT when he began to look at similar equations with independent exponents. He constructed several algorithms to generate solution sets but the very nature of the algorithms he was able to construct required a common factor in the bases. He began to suspect that co-prime bases might be impossible and set out to test his hypothesis by computer. Andy Beal and a colleague programmed 15 computers and after thousands of cumulative hours of operation had checked all variable values through 99. Many solutions were found: all had a common factor in the bases. While certainly not conclusive, Andy Beal now had sufficient reason to share his discovery with the world.*

***BEAL'S CONJECTURE:*** *If $A^x + B^y = C^z$, where A, B, C, x, y and z are positive integers and x, y and z are all greater than 2, then A, B and C must have a common prime factor.*

*Andy Beal wrote many letters to mathematics periodicals and number theorists. Among the replies were two considered responses from number theorists. Dr. Harold Edwards from the department of mathematics at New York University and author of "Fermat's Last Theorem, a genetic introduction to algebraic number theory" confirmed that the discovery was unknown and called it "quite remarkable". Dr. Earl Taft from the department of mathematics at Rutgers University relayed Andy Beal's discovery to Jarell Tunnell who was "an expert on Fermat's Last Theorem", according to Taft's response, and also confirmed that the discovery and conjecture were unknown. There is no known evidence of prior knowledge of Beal's conjecture and all references to it begin after Andy Beal's 1993 discovery and subsequent dissemination of it. The related ABC conjecture hypothesizes that only a finite number of solutions could exist.*

## *Interesting Sidenote*

*While Beal's conjecture was widely received with enthusiasm by the mathematics community at large, it seems that there are always people with other motivations.*

*Three mathematicians by the names of Andrew Granville, Alf Vanderpoorten, and Andrew Bremmer set out to discredit and criticize Andy Beal and the attribution of the conjecture, including writing scathing criticisms claiming that the conjecture was previously known and evidenced in prior works dating back to Brun's work of 1914. The untruthfulness and motivations of their criticisms were made obvious by the subsequent 1996 publication of Vanderpoorten's book wherein he claims to propose the conjecture as recent original thought by none other than . . . himself (see page 194 of Vanderpoorten's "notes on Fermat's Last Theorem," 1st edition).*

## *Encouraging Others*

*By offering a cash prize for the proof or disproof of this important number theory relationship, Andy Beal hopes to inspire young minds to think about the equation, think about winning the offered prize, and in the process become more interested in the wonderful study of mathematics. Information regarding the $1,000,000 cash prize that is held in trust by the American Mathematics Society can be obtained at the University of North Texas web site: http://www.math.unt.edu/~mauldin/beal.html.*

*Incidentally, Andy Beal believes that the world has yet to adequately respond to Fermat's challenge to the English in 1657 regarding Pell's equation (see Edwards book referenced above - book section 1.9). Andy Beal believes that Fermat may well have had a method of solution that was "not inferior to the more celebrated questions of geometry in respect of beauty, difficulty, or method of proof". Fermat claimed his method involved infinite descent and no known methods of solution use a descent. Furthermore, the continued fraction and cyclic methods of solution known today hardly qualify as beautiful or particularly difficult. For someone seeking to learn about or advance number theory, that's a great and fun place to start.*

*The Beal Conjecture is sometimes referred to as "Beal's conjecture", "Beal's problem" or the "Beal problem".*

# PROVE

Andy Beal's Conjecture derived from Fermat's last theorem, of course very difficult, if easy, the mathematicians have solved it. But with my popular method, we can be proved,

We read it several times to record the conditions for such:

- "*If* $A^x + B^y = C^z$"
- "*A, B, C, x, y and z are positive integers*"
- "*x, y and z are all greater than 2,*"
- "*then A, B and C must have a common prime factor*"

Back to Andy Beal's Conjecture with equation

$$A^x + B^y = C^z$$

This Diophantine Equation is conditional, it is very difficult for us, the conditions A, B, C positive integers, the exponent x, y, z, positive integer too, and greater than 2.

My popular method is very simple, I choose any value of A, B, x, y positive integers and according to the conditions of the Beal conjecture.

I now prove that C and z have the same conditions of Andy Beal Conjecture

Choose:

$A \equiv \text{mod } Z$

$B \equiv \text{mod } Z$ or $B \equiv \text{mod } A$

From this equation $A^x + B^y = C^z$

$$C = \sqrt[z]{A^x + B^y}$$

I prove that C is common A or P factor ?

$$C = \sqrt[z]{A^x + B^y}$$

$$C = \sqrt[z]{A^x} \cdot \sqrt[z]{1 + (B^y / A^x)}$$

Special cases of the Beal Conjecture

$x = y = z = n$

This equation $A^x + B^y = C^z$

To $A^n + B^n = C^n$

$$C = \sqrt[n]{A^n} \cdot \sqrt[n]{1 + (B^n / A^n)}$$

$$C \equiv A \cdot \sqrt[n]{1+modZ^{n}}$$

n > 2 this is Fermat's Last Theorem

$$C \equiv A \cdot \sqrt[n]{1+modZ^{n}}$$

$\sqrt[n]{1+modZ^{n}}$ was irrational

A, B are positive integers then C is not positive intege

1995 Mathematics Conference at Boston University recognized by Professor Andrew Wiles proved FLT, therefore we do not need to prove Fermat's Last Theorem

Or see the Fermat's Last Theorem by Ran Van Vo, published in 2002,

Back to: B ≡ mod Z or B ≡ modA

And $y \neq x$

$B^y/A^x \equiv modZ^{|y-x|}$

$$C = \sqrt[z]{A^x} \cdot \sqrt[z]{1+\left(\frac{B^y}{A^x}\right)} \equiv \sqrt[z]{A^x} \cdot \sqrt[z]{1+modZ^{|y-x|}}$$

$\sqrt[z]{A^x}$ : A, x, positive integers and according to the conditions of the Beal conjecture,

$x \equiv modz$

$$C \equiv \sqrt[z]{A^{modz}} \cdot \sqrt[z]{1+modZ^{|y-x|}}$$

$$C \equiv A \cdot \sqrt[z]{1+modZ^{|y-x|}}$$

$\sqrt[z]{1+modZ^{|y-x|}}$ zth root perfect

Then C must have a common A or Prime factor

Reality

$$3^3 + 6^3 = 3^5$$

$$3^9 + 54^3 = 3^{11}$$

$$3^6 + 18^3 = 3^8$$

CONCLUDE

We can conclude that:

*"If A^x + B^y = C^ z, where A, B, C, x, y and z are positive integers and x, y and z are all greater than 2, then A, B and C must have a common prime factor"* is TRUE

# NEW FORMULA

From the Beal conjecture I have a new Method, can say that it's a new formula

If we have A, B, positive integers and according to the conditions of the Beal conjecture, we find the value of C by the following formula:

$$(A,B,C) \in Z$$

$$(x,y,z) \in Z > 2$$

$$C \equiv \sqrt[z]{A^{modz}} \cdot \sqrt[z]{1+modZ^{|y-x|}}$$

$$\rightarrow gcf(A,B,C) \geq 2$$

Ex:

1*) Find the values of A, B, C are positive integers, which must have a common prime factor and the exponent x, y, z are all greater than 2

$$A^x + B^y = C^z$$

Prove

We have Diophantine Equation

$$A^x + B^y = C^z$$

MY METHOD

$$(A,B,C) \in Z$$

$$(x,y,z) \in Z > 2$$

$$C \equiv \sqrt[z]{A^{modz}} \cdot \sqrt[z]{1 + modZ^{|y-x|}}$$

$$\rightarrow gcf(A,B,C) \geq 2$$

** Choose any value of the exponent z (z > 2)
3, 4, 5, 6, 7, 8, 9 . . . . (i.e., z = 9)

** Choose Z any positive integer (Z > 1)
2, 3, 4, 5, 7, . . . (i.e., Z = 7)

** We have $Z^z - 1 = 7^9 - 1 = 40353606$

Following new formula:

$$(A,B,C) \in \mathbf{Z}$$

$$(x,y,z) \in \mathbf{Z} > 2$$

$$C \equiv \sqrt[z]{A^{modz}}\ \sqrt[z]{1+modZ^{|y-x|}}$$

$$\rightarrow gcf(A,B,C) \geq 2 \ or \ P$$

** The values of A, B, C

B ≡ mod A, or B = A = 40353606

And

C = Z.A = 7 40353606 = 282475242

Then the values of A, B, C

40353606 + 40353606 = 282475242

** The exponent z = 9

x ≡ mod z, or x = z = 9

We choose |y-x| = 1 → y = x + 1 = 9 + 1 = 10

Exponent: 9, 10, 9

** Results

$40353606^9 + 40353606^{10} = 282475242^9$

We rewrite

$(40353606^3)^3 + 40353606^{10} = 282475242^9$

$6571235747829360746901 6^3 + 40353606^{10} = 282475242^9$

Try again

Left side of the equation

$65712357478293607469016^3 + 40353606^{10} =$
$1.1450475040539714879582358976006e+76$

Right side of the equation

$282475242^9 = 1.1450475040539714879582358976006e+76$

** Solution

$$A^x + B^y = C^z$$

$65712357478293607469016^3 + 40353606^{10} = 282475242^9$

x, y and z are all greater than 2
A, B and C have common factor for 3

A = 65712357478293607469016, x = 3
B = 40353606, y = 10
C = 282475242, z = 9

2*) Find the values of A, B, C are positive integers, which must have a common prime factor and the exponent x, y, z are all greater than 2

$$A^x + B^y = C^z$$

Prove

We have Diophantine Equation

$$A^x + B^y = C^z$$

Similarly we use the new formula

$$(A, B, C) \in Z$$

$$(x, y, z) \in Z > 2$$

$$C \equiv \sqrt[z]{A^{modz}} \cdot \sqrt[z]{1 + modZ^{|y-x|}}$$

$$\rightarrow gcf(A, B, C) \geq 2$$

** Choose any value of the exponent z ($z > 2$)
3, 4, 5, 6, 7, 8, 9 . . . . (i.e., z = 17)

** Choose Z any positive integer ($Z > 1$)
2, 3, 4, 5, 7, . . . (i.e., Z = 2)

** We have $Z^z - 1 = 2^{17} - 1 = 131071$

$$(A, B, C) \in \boldsymbol{Z}$$

$$(x, y, z) \in \boldsymbol{Z} > 2$$

$$C \equiv \sqrt[z]{A^{modz}} \; \sqrt[z]{1 + modZ^{|y-x|}}$$

$$\rightarrow gcf(A, B, C) \geq 2 \text{ or } P$$

** The values of A, B, C

A ≡ mod Z, or A = 131071

B ≡ mod Z, or B = 131071

And

C = Z.A = 2 131071 = 262142

Then the values of A, B, C

131071 + 131071 = 262142

** The exponent z = 17

Then

x ≡ mod z, or x = z = 17

We choose |y-x| = 1 → y = x + 1 = 17 + 1 = 18

Exponent: 17, 18, 17

** Results

$131071^{17} + 131071^{18} = 262142^{17}$

We rewrite

$131071^{17} + (131071^{2})^{9} = 262142^{17}$

$131071^{17} + 17179607041^{9} = 262142^{17}$

Try again

Left side of the equation

$131071^{17} + 17179607041^{9}$
$= 1.3035339452737265751117827282387e+92$

And right side of the equation

$262142^{17} = 1.3035339452737265751117827282387e+92$

A, B, C common 131071 factor

** Solution

| | |
|---|---|
| A = 131071 | x = 17 |
| B = 17179607041 | y = 9 |
| C = 262142 | z = 17 |

3*) Find the values of A, B, C are positive integers, which must have a common prime factor and the exponent x, y, z are all greater than 2

$$A^x + B^y = C^z$$

Prove

The Diophantine Equation:

$$A^x + B^y = C^z$$

Similarly we use the new formula

$$(A,B,C) \in \boldsymbol{Z}$$

$$(x,y,z) \in \boldsymbol{Z} > 2$$

$$C \equiv \sqrt[z]{A^{modz}}\ \sqrt[z]{1+modZ^{|y-x|}}$$

$$\rightarrow gcf(A,B,C) \geq 2 \ or \ P$$

** Choose any value of the exponent z (z > 2)
3, 4, 5, 6, 7, 8, 9 . . . . (i.e., z = 9)

** Choose Z any positive integer (Z > 1)
2, 3, 4, 5, 7, . . . (i.e., Z = 4)

** We have $Z^z$ -1 = $4^9$ – 1 = 262143

$$(A,B,C) \in \boldsymbol{Z}$$

$$(x,y,z) \in \boldsymbol{Z} > 2$$

$$C \equiv \sqrt[z]{A^{modz}}\ \sqrt[z]{1+modZ^{|y-x|}}$$

$$\rightarrow gcf(A,B,C) \geq 2 \ or \ P$$

** The values of A, B, C

A ≡ mod Z, or A = 262143

B ≡ mod Z, or B = 262143

And

C = Z.A = 4 262143 = 1048572

Then the values of A, B, C

262143 + 262143 = 1048572

** The exponent z = 9

Then

$x \equiv \text{mod } z$, or $x = z = 9$

We choose $|y-x| = 1$ $y = x + 1 = 9 + 1 = 10$

Exponent: 9, 10, 9

** Results

$262143^9 + 262143^{10} = 1048572^9$

We rewrite

$(262143^3)^3 + (262142^2)^5 = 1048572^9$

$18014192351838207^3 + 68718952449^5 = 1048572^9$

Try again

Left side of the equation

$18014192351838207^3 + 68718952449^5$
$= 1.5324429276097646271115779696648e+54$

And right side of the equation

$1048572^9 = 1.5324429276097646271115779696648e+54$

** Solution:

A = 18014192351838207 x = 3
B = 68718952449 y = 5
C = 1048572 z = 9

4*) Find the values of A, B, C, they *must have a common prime factor* and the exponents x, y, z are greater than 2

$$A^x + B^y = C^z$$

Prove

We have Diophantine Equation

$$A^x + B^y = C^z$$

Similarly we use the new formula

$$(A,B,C) \in \mathbf{Z}$$

$$(x,y,z) \in \mathbf{Z} > 2$$

$$C \equiv \sqrt[z]{A^{modz}}\ \sqrt[z]{1+modZ^{|y-x|}}$$

$$\rightarrow gcf(A,B,C) \geq 2 \text{ or } P$$

** Choose any value of the exponent z ($z > 2$)
3, 4, 5, 6, 7, 8, 9 . . . . (i.e., $z = 12$)

** Choose Z any positive integer ($Z > 1$)
2, 3, 4, 5, 7, . . . (i.e., $Z = 5$)

** We have $Z^z - 1 = 5^{12} - 1 = 244140624$

$$(A,B,C) \in \boldsymbol{Z}$$

$$(x,y,z) \in \boldsymbol{Z} > 2$$

$$C \equiv \sqrt[z]{A^{modz}} \; \sqrt[z]{1 + modZ^{|y-x|}}$$

$$\rightarrow gcf(A,B,C) \geq 2 \; or \; P$$

** The values of A, B, C

$A \equiv$ mod Z, or $A = 244140624$

$B \equiv$ mod Z, or $B = 244140624$

And

$C = Z.A = 5\ 244140624 = 1220703120$

Then the values of A, B, C

$244140624 + 244140624 = 1220703120$

** The exponent $z = 12$

Then

$x \equiv \mod z$, or $x = z = 12$

We choose $|y-x| = 1 \rightarrow y = x + 1 = 12 + 1 = 13$

Exponent: 12, 13, 12

** Results

$244140624^{12} + 244140624^{13} = 1220703120^{12}$

We rewrite

$(244140624^{2})^{6} + 244140624^{13} = 1220703120^{12}$

$59604644287109376^{6} + 244140624^{13} = 1220703120^{12}$

Try again

Left equation

$59604644287109376^{6} + 244140624^{13}$
$= 1.0947643714439035188147209479977e+109$

Right equation

$1220703120^{12} = 1.0947643714439035188147209479977e+109$

** Solution:

$A = 59604644287109376$,  $x = 6$

B = 244140624 y = 13
C = 1220703120 z = 12

There are the Diophantine Equations have form Beal Equation "$A^x + B^y = C^z$" we use new formula:

$$(A,B,C) \in \boldsymbol{Z}$$

$$(x,y,z) \in \boldsymbol{Z} > 2$$

$$C \equiv \sqrt[z]{A^{modz}}\ \sqrt[z]{1+modZ^{|y-x|}}$$

$$\rightarrow gcf(A,B,C) \geq 2 \text{ or } P$$

Ex:

$$A^x + B^y = C^z$$

We have a lot of results like this:

$$35181150961663^5 + 1073676289^8 = 65534^{15}$$

. . . .

Try again

Left equation

$$35181150961663^5 + 1073676289^8$$
$$= 1.7660384391335017705152603622735e+72$$

Right equation

$$65534^{15} = 1.7660384391335017705152603622735e+72$$

A, B, C common 7 (prime number) factor

Solution

A = 35181150961663 x = 5
B = 1073676289 y = 8
C = 65534 z = 15

. . . .

# EXERCISE

We have a new formula above, step and step, we find the value A, B, C, x, y, z, where A, B, C, x, y and z are positive integers and x, y and z are all greater than 2, then A, B and C must have a common prime factor, of the form of Beal Conjecture

$$A^x + B^y = C^z$$

$$(A,B,C) \in \boldsymbol{Z}$$

$$(x,y,z) \in \boldsymbol{Z} > 2$$

$$C \equiv \sqrt[z]{A^{modz}} \; \sqrt[z]{1 + modZ^{|y-x|}}$$

$$\rightarrow gcf(A,B,C) \geq 2 \; or \; P$$

1*) Find the values of A, B, C, x, y, z (all positive integers) where exponents x, y, z are greater than 2, of the following Beal Equation

$$A^x + B^y = C^z$$

Given z = 4; and {gcf(A,B,C) = P}

2*) Find the values of A, B, C, x, y, z (all positive integers) where exponents x, y, z are greater than 2, of the following Beal Equation

$$A^x + B^y = C^z$$

Given z = 7; and {gcf(A,B,C) = P}

3*) Find the values of A, B, C, x, y, z (all positive integers) where exponents x, y, z are greater than 2, of the following Beal Equation

$$A^x + B^y = C^z$$

Given z = 8; and {gcf(A,B,C) = P}

4*) Find the values of A, B, C, x, y, z (all positive integers) where exponents x, y, z are greater than 2, of the following Beal Equation

$$A^x + B^y = C^z$$

Given A = 262143; and {gcf(A,B,C) = P}

5*) Find the values of A, B, C, x, y, z (all positive integers) where exponents x, y, z are greater than 2, of the following Beal Equation

$$A^x + B^y = C^z$$

Given C = 177144; and {gcf(A,B,C) = P}

6*) Find the values of A, B, C, x, y, z (all positive integers) where exponents x, y, z are greater than 2, of the following Beal Equation

$$A^x + B^y = C^z$$

Given z = 21; and {gcf(A,B,C) = P}

7*) Find the values of A, B, C, x, y, z (all positive integers) where exponents x, y, z are greater than 2, of the following Beal Equation

$$A^x + B^y = C^z$$

Given z = 11; and {gcf(A,B,C) = P}

8*) Find the values of A, B, C, x, y, z (all positive integers) where exponents x, y, z are greater than 2, of the following Beal Equation

$$A^x + B^y = C^z$$

Given z = 12; and {gcf(A,B,C) = P}

9*) Find the values of A, B, C, x, y, z (all positive integers) where exponents x, y, z are greater than 2, of the following Beal Equation

$$A^x + B^y = C^z$$

Given z = 13; and {gcf(A,B,C) = P}

10*) Find the values of A, B, C, x, y, z (all positive integers) where exponents x, y, z are greater than 2, of the following Beal Equation

$$A^x + B^y = C^z$$

Given z = 14; and {gcf(A,B,C) = P}

11*) Find the values of A, B, C, x, y, z (all positive integers) where exponents x, y, z are greater than 2, of the following Beal Equation

$$A^x + B^y = C^z$$

Given z = 15; and {gcf(A,B,C) = P}

12*) Find the values of A, B, C, x, y, z (all positive integers) where exponents x, y, z are greater than 2, of the following Beal Equation

$$A^x + B^y = C^z$$

Given z = 16; and {gcf(A,B,C) = P}

13*) Find the values of A, B, C, x, y, z (all positive integers) where exponents x, y, z are greater than 2, of the following Beal Equation

$$A^x + B^y = C^z$$

Given z = 17; and {gcf(A,B,C) = P}

# EXPANSION THE BEAL CONJECTURE

$$A^v + B^x + C^y = D^z$$

EXPANSION

We expand many times on the same form of the equation $A^x + B^y = C^z$, so I extended over many unknowns for the Beal Equation:

Ex:

$$A^v + B^x + C^y = D^z$$

$$A^u + B^v + C^x + D^y = E^z$$

$$A^t + B^u + C^v + D^x + E^y = F^z$$

. . . .

$A \equiv \bmod\delta, (\delta > 1)$
$B \equiv \bmod\delta$
$C \equiv \bmod\delta$

And

$$x \equiv \text{mod}z\ (x \geq z)$$

Δ: left side of the unknowns of the equation

And $|y - x| \geq 1$

We apply general method above for the form Equation, which expand to many unknowns,

$$A^v + B^x + C^y = D^z$$

$$A^u + B^v + C^x + D^y = E^z$$

. . . .

$$(A,B,C) \in \boldsymbol{Z}$$

$$(x,y,z) \in \boldsymbol{Z} > 2$$

$$C \equiv \sqrt[z]{A^{modz}}\ \sqrt[z]{(\Delta - 1) + modZ^{|y-x|}}$$

Δ: *left term's Equation*

$$\rightarrow gcf(A,B,C) \geq 2$$

There are two objectives, when we extent the Beal conjecture

1*) First objective: the Beal conjecture was also true for the extended equation

$$A^v + B^x + C^y = D^z$$

2*) The second objective: the General method is good solution for the extended equation

$$(A,B,C) \in \boldsymbol{Z}$$

$$(x,y,z) \in \boldsymbol{Z} > 2$$

$$C \equiv \sqrt[z]{A^{modz}} \; \sqrt[z]{(\Delta-1)+modZ^{|y-x|}}$$

$\Delta$: *left term's Equation*

$$\rightarrow gcf(A,B,C) \geq 2$$

***To the following problem***

5*) Find the values of A, B, C, D, v, x, y, z (all positive integers) where exponents v, x, y, z are greater than 2, of the following Beal Equation

$$A^v + B^x + C^y = D^z$$

Prove

The Diophantine Equation has form the Beal conjecture

$$A^v + B^x + C^y = D^z$$

Applying the new General method

$$(A,B,C) \in \boldsymbol{Z}$$

$$(x,y,z) \in \boldsymbol{Z} > 2$$

$$C \equiv \sqrt[z]{A^{modz}} \sqrt[z]{(\Delta - 1) + modZ^{|y-x|}}$$

$\Delta$: *left term's Equation*

$$\rightarrow gcf(A,B,C) \geq 2$$

Similarly above

** Choose any value of the exponent z (z > 2)
3, 4, 5, 6, 7, 8, 9 . . . . (i.e., z = 8)

** Choose Z any positive integer (Z > 1)
2, 3, 4, 5, 7, . . . (i.e., Z = 6)

$(\Delta - 1) = 3\text{-}1 = 2$ ($\Delta$=3 *left term's Equation*)

** We have $Z^z - 2 = 6^8 - 2 = 1679614$

** The values of A, B, C

$A \equiv mod\delta = 1679614$
$B \equiv mod\delta = 1679614$
$C \equiv mod\delta = 1679614$

And

D = Z.A = 6 1679614 = 10077684

Then the values of A, B, C, D

1679614 + 1679614 + 1679614 = 10077684

** The exponent z = 8

Then v and x equal to

x ≡ mod z, or x = z = 8

We choose |y-x| = 1 → y = x + 1 = 8 + 1 = 9

We have the exponents: 8, 8, 9, 8

** Results

$1679614^8 + 1679614^8 + 1679614^9 = 10077684^8$

We rewrite

$(1679614^2)^4 + 1679614^8 + (1679614^3)^3 = 10077684^8$

$2821103188996^4 + 1679614^8 + 4738364411682327544^3 = 10077684^8$

Try again

Left side of the equation

$$2821103188996^4 + 1679614^8 + 4738364411682327544^3 = 1.0638634548335355836019005142614e+56$$

Right side of the equation

$$10077684^8 = 1.0638634548335355836019005142614e+56$$

** Solution:

A = 2821103188996, v = 4
B = 1679614 x = 8
C = 4738364411682327544 y = 3
D = 10077684 z = 8

6*) Find the values of A, B, C, D, v, x, y, z (all positive integers) where exponents v, x, y, z are greater than 2, of the following Beal Equation

$$A^v + B^x + C^y = D^z$$

v,x,y,z > 2; and {gcf(A,B,C,D) = P}

Prove

Similarly, applying the new formula

$$(A,B,C) \in \boldsymbol{Z}$$

$$(x,y,z) \in \boldsymbol{Z} > 2$$

$$C \equiv \sqrt[z]{A^{modz}} \; \sqrt[z]{(\Delta - 1) + modZ^{|y-x|}}$$

$\Delta$: *left term's Equation*

$\rightarrow gcf(A,B,C) \geq 2$

We have

$$295431147120421331762^{5} + 14348905^{15} + 205891074699025^{8} = 43046715^{15}$$

Try again

Left equation

$$295431147120421331762^{5} + 14348905^{15} + 205891074699025^{8} = 3.229239266453922887064595919856 1e+114$$

Right equation

$$43046715^{15} = 3.2292392664539228870645959198561e+114$$

gcf(A, B, C, D) = 5

and exponents v,x,y,z are greater than 2

Solution

A = 295431147120421331762 5      v = 5
B = 14348905      x = 15
C = 205891074699025      y = 8
D = 43046715      z = 15

. . . .

7*) Find the values of A, B, C, D, v, x, y, z (all positive integers) where exponents v, x, y, z are greater than 2, of the following Beal Equation

$$A^v + B^x + C^y = D^z$$

v,x,y,z > 2; and

$$\{gcf(A,B,C,D) = P\}$$

Prove

Applying the new formula

$$(A,B,C) \in \mathbf{Z}$$

$$(x,y,z) \in \mathbf{Z} > 2$$

$$C \equiv \sqrt[z]{A^{modz}} \; \sqrt[z]{(\Delta - 1) + modZ^{|y-x|}}$$

$\Delta$: *left term's Equation*

$$\rightarrow gcf(A,B,C) \geq 2$$

We have

$$542800770374277320431105441^6 + 23298085122479^{12} + 23298085122479^{13} = 302875106592227^{12}$$

Try again

Left Equation

$$542800770374277320431105441^{6} + 23298085122479^{12} + 23298085122479^{13}$$
$$= 5.9588569515359096021133219195012e{+}173$$

Right equation

$$302875106592227^{12}$$
$$= 5.9588569515359096021133219195012e{+}173$$

gcf(A, B, C, D) = 23298085122479

And exponents v,x,y,z are greater than 2

Solution

A = 542800770374277320431105441 v = 6
B = 23298085122479 x = 12
C = 23298085122479 y = 13
D = 302875106592227 z = 12

. . . .

# EXERCISE

*) Find the values of A, B, C, D, v, x, y, z (all positive integers) where exponents v, x, y, z are greater than 2, of the following Beal Equation

$$A^v + B^x + C^y = D^z$$

Given v = 5 and

$$\{gcf(A,B,C,D) = P\}$$

*) Find the values of A, B, C, D, v, x, y, z (all positive integers) where exponents v, x, y, z are greater than 2, of the following Beal Equation

$$A^v + B^x + C^y = D^z$$

Given z = 7 and v,x,y > 2;

$$\{gcf(A,B,C,D) = P\}$$

*) Find the values of A, B, C, D, v, x, y, z (all positive integers) where exponents v, x, y, z are greater than 2, of the following Beal Equation

$$A^v + B^x + C^y = D^z$$

Given y = 10 and v,x,z > 2;

$$\{gcf(A,B,C,D) = P\}$$

*) Find the values of A, B, C, D, v, x, y, z (all positive integers) where exponents v, x, y, z are greater than 2, of the following Beal Equation

$$A^v + B^x + C^y = D^z$$

Given z = 9, v,x,y > 2; and

$$\{gcf(A,B,C,D) = P\}$$

*) Find the values of A, B, C, D, v, x, y, z (all positive integers) where exponents v, x, y, z are greater than 2, of the following Beal Equation

$$A^v + B^x + C^y = D^z$$

Given z = 15, v,x,y, > 2; and

$$\{gcf(A,B,C,D) = P\}$$

*) Find the values of A, B, C, D, v, x, y, z (all positive integers) where exponents v, x, y, z are greater than 2, of the following Beal Equation

$$A^v + B^x + C^y = D^z$$

Given v = 3, x,y,z > 2; and

$\{gcf(A,B,C,D) = P\}$

*) Find the values of A, B, C, D, v, x, y, z (all positive integers) where exponents v, x, y, z are greater than 2, of the following Beal Equation

$A^v + B^x + C^y = D^z$

Given $x = 5$, $v,y,z > 2$; and

$\{gcf(A,B,C,D) = P\}$

*) Find the values of A, B, C, D, v, x, y, z (all positive integers) where exponents v, x, y, z are greater than 2, of the following Beal Equation

$A^v + B^x + C^y = D^z$

Given $z = 7$, $v,x,y > 2$; and

$\{gcf(A,B,C,D) = P\}$

*) Find the values of A, B, C, D, v, x, y, z (all positive integers) where exponents v, x, y, z are greater than 2, of the following Beal Equation

$A^v + B^x + C^y = D^z$

Given $v,x,y,z > 2$; and

$\{gcf(A,B,C,D) = P\}$

*) Find the values of A, B, C, D, v, x, y, z (all positive integers) where exponents v, x, y, z are greater than 2, of the following Beal Equation

$A^v + B^x + C^y = D^z$

Given y = 6, v,x,z > 2; and

$$\{gcf(A,B,C,D) = P\}$$

*) Find the values of A, B, C, D, v, x, y, z (all positive integers) where exponents v, x, y, z are greater than 2, of the following Beal Equation

$A^v + B^x + C^y = D^z$

Given v,x,y,z > 2; and

$$\{gcf(A,B,C,D) = P\}$$

*) Find the values of A, B, C, D, v, x, y, z (all positive integers) where exponents v, x, y, z are greater than 2, of the following Beal Equation

$A^v + B^x + C^y = D^z$

Given z = 11 v,x,y > 2; and

$$\{gcf(A,B,C,D) = P\}$$

*) Find the values of A, B, C, D, v, x, y, z (all positive integers) where exponents v, x, y, z are greater than 2, of the following Beal Equation

$$A^v + B^x + C^y = D^z$$

Given $z = 21$, $v,x,y, > 2$; and

$$\{gcf(A,B,C,D) = P\}$$

*) Find the values of A, B, C, D, v, x, y, z (all positive integers) where exponents v, x, y, z are greater than 2, of the following Beal Equation

$$A^v + B^x + C^y = D^z$$

Given $z = 12$, $v,x,y > 2$; and

$$\{gcf(A,B,C,D) = P\}$$

# EXPANSION OF THE BEAL CONJECTURE TO FORM

$$A^u + B^v + C^x + D^y = E^z$$

Similarly

There are two objectives, when we extent the Beal conjecture

1) First objective: the Beal conjecture was also true for the extended equation

$$A^u + B^v + C^x + D^y = E^z$$

2) The second objective: the General method is good solution for the extended equation

$$(A,B,C) \in \boldsymbol{Z}$$

$$(x,y,z) \in \boldsymbol{Z} > 2$$

$$C \equiv \sqrt[z]{A^{modz}} \; \sqrt[z]{(\Delta - 1) + modZ^{|y-x|}}$$

Δ: *left term's Equation*

$\rightarrow gcf(A,B,C) \geq 2$

***To the following problem***

*) Find the values of A, B, C, D, E, u, v, x, y, z (all positive integers) where exponents u, v, x, y, z are greater than 2, of the following Beal Equation

$$A^u + B^v + C^x + D^y = E^z$$

u,v,x,y,z > 2; and

$$\{gcf(A,B,C,D,E) = P\}$$

Prove

Applying the new formula

$(A,B,C) \in \boldsymbol{Z}$

$(x,y,z) \in \boldsymbol{Z} > 2$

$$C \equiv \sqrt[z]{A^{modz}} \; \sqrt[z]{(\Delta - 1) + modZ^{|y-x|}}$$

Δ: *left term's Equation*

$\rightarrow gcf(A,B,C) \geq 2$

Similarly above

** Choose any value of the exponent z ($z > 2$)
3, 4, 5, 6, 7, 8, 9, . . . . (i.e., $z = 12$)

** Choose Z any positive integer ($Z > 1$)
2, 3, 4, 5, 7, 8, 9, . . . (i.e., $Z = 2$)

$(\Delta - 1) = 4\text{-}1 = 3$ ($\Delta = 4$ *left term's Equation*)

** we have $Z^z - 3 = 2^{12} - 3 = 4093$

** The values of A, B, C, D

$A \equiv \text{mod}\delta = 4093$
$B \equiv \text{mod}\delta = 4093$
$C \equiv \text{mod}\delta = 4093$
$D \equiv \text{mod}\delta = 4093$

And

$E = Z.A = 2\ 4093 = 8186$

Then we have the values of A, B, C, D, E

$4093 + 4093 + 4093 + 4093 = 8186$

** The exponent $z = 12$

Then u, v and x equal to

$x \equiv \text{mod } z$, or $x = z = 12$

We choose $|y-x| = 1$ $y = x + 1 = 12 + 1 = 13$

We have the exponents: 12, 12, 12, 13, 12

** Results

$4093^{12} + 4093^{12} + 4093^{12} + 4093^{13} = 8186^{12}$

Rewrite

$(4093^{2})^{6} + (4093^{3})^{4} + (4093^{4})^{3} + 4093^{13} = 8186^{12}$

We have

$$16752649^{6} + 68568592357^{4} + 4093^{13} + 280651248517201^{3} = 8186^{12}$$

Try again

Left equation

$$16752649^{6} + 68568592357^{4} + 4093^{13} + 280651248517201^{3} = 9.05442516670315880861469168741 64e+46$$

And right equation

$$8186^{12} = 9.0544251667031588086146916874164e+46$$

Solution

| | |
|---|---|
| A = 16752649 | u = 6 |
| B = 68568592357 | v = 4 |
| C = 4093 | x = 13 |

D = 280651248517201 y = 3
E = 8186 z = 12

*) Find the values of A, B, C, D, E, u, v, x, y, z (all positive integers) where exponents u, v, x, y, z are greater than 2, of the following Beal Equation

$$A^u + B^v + C^x + D^y = E^z$$

u,v,x,y,z > 2; and

{gcf(A,B,C,D,E) = P}

Prove

Applying the new formula

$$(A, B, C) \in \boldsymbol{Z}$$

$$(x, y, z) \in \boldsymbol{Z} > 2$$

$$C \equiv \sqrt[z]{A^{modz}} \sqrt[z]{(\Delta - 1) + modZ^{|y-x|}}$$

Δ: *left term's Equation*

$$\rightarrow gcf(A, B, C) \geq 2$$

Similarly above

** Choose any value of the exponent z (z > 2)
3, 4, 5, 6, 7, 8, 9, . . . . (i.e., z = 18)

** Choose Z any positive integer ($Z > 1$)
2, 3, 4, 5, 7, 8, 9, . . . (i.e., $Z = 3$)

$(\Delta - 1) = 4\text{-}1 = 3$ ($\Delta$ =*4 left term's Equation*)

** we have $Z^z - 3 = 3^{18} - 3 = 387420486$

** The values of A, B, C, D

$A \equiv \text{mod}\delta = 387420486$
$B \equiv \text{mod}\delta = 387420486$
$C \equiv \text{mod}\delta = 387420486$
$D \equiv \text{mod}\delta = 387420486$

And

$E = Z.A = 3\ 387420486 = 1162261458$

Then we have the values of A, B, C, D, E

$387420486 + 387420486 + 387420486 + 387420486 = 1162261458$

** The exponent $z = 18$

Then u, v and x equal to

$x \equiv \text{mod } z$, or $x = z = 18$

We choose $|y\text{-}x| = 1 \rightarrow y = x + 1 = 18 + 1 = 19$

We have the exponents: 18, 18, 18, 19, 18

** Results

$$387420486^{18}+387420486^{18}+387420486^{18}+387420486^{19} = 1162261458^{18}$$

Rewrite

$$(387420486^{2})^{9}+(387420486^{3})^{6}+387420486^{18}+387420486^{19} = 1162261458^{18}$$

We have

$$150094632972476196^{9}+58149735652188352477751256^{6} +387420486^{18}+387420486^{19} = 1162261458^{18}$$

Try again
Left equation

$$150094632972476196^{9}+58149735652188352477751256^{6} +387420486^{18}+387420486^{19} = 1.4978525171549833800261368278585e+163$$

And right equation

$$1162261458^{18} = 1.4978525171549833800261368278585e+163$$

Solution

A = 150094632972476196 u = 9
B = 58149735652188352477751256 v = 6
C = 387420486 x = 18
D = 387420486 y = 19
E = 1162261458 z = 18

# EXERCISE

*) Find the values of A, B, C, D, E, u, v, x, y, z (all positive integers) where exponents u, v, x, y, z are greater than 2, of the following Beal Equation

$$A^u + B^v + C^x + D^y = E^z$$

Given u = 6, v,x,y,z > 2; and

$$\{gcf(A,B,C,D,E) = P\}$$

*) Find the values of A, B, C, D, E, u, v, x, y, z (all positive integers) where exponents u, v, x, y, z are greater than 2, of the following Beal Equation

$$A^u + B^v + C^x + D^y = E^z$$

Given z = 8 u,v,x,y > 2; and

$$\{gcf(A,B,C,D,E) = P\}$$

*) Find the values of A, B, C, D, E, u, v, x, y, z (all positive integers) where exponents u, v, x, y, z are greater than 2, of the following Beal Equation

$$A^u + B^v + C^x + D^y = E^z$$

Given u,v,x,y,z > 2; A = 262141, and

{gcf(A,B,C,D,E) = P}

*) Find the values of A, B, C, D, E, u, v, x, y, z (all positive integers) where exponents u, v, x, y, z are greater than 2, of the following Beal Equation

$$A^u + B^v + C^x + D^y = E^z$$

Given u,v,x,y,z > 2; and

{gcf(A,B,C,D,E) = P}

*) Find the values of A, B, C, D, E, u, v, x, y, z (all positive integers) where exponents u, v, x, y, z are greater than 2, of the following Beal Equation

$$A^u + B^v + C^x + D^y = E^z$$

Given u,v,x,y,z > 2; and

{gcf(A,B,C,D,E) = P}

*) Find the values of A, B, C, D, E, u, v, x, y, z (all positive integers) where exponents u, v, x, y, z are greater than 2, of the following Beal Equation

$$A^u + B^v + C^x + D^y = E^z$$

Given z = 11; and

$\{gcf(A,B,C,D,E) = P\}$

*) Find the values of A, B, C, D, E, u, v, x, y, z (all positive integers) where exponents u, v, x, y, z are greater than 2, of the following Beal Equation

$A^u + B^v + C^x + D^y = E^z$

Given z = 7 ; and

$\{gcf(A,B,C,D,E) = P\}$

*) Find the values of A, B, C, D, E, u, v, x, y, z (all positive integers) where exponents u, v, x, y, z are greater than 2, of the following Beal Equation

$A^u + B^v + C^x + D^y = E^z$

Given u,v,x,y,z > 2; A = 262141, and

$\{gcf(A,B,C,D,E) = P\}$

*) Find the values of A, B, C, D, E, u, v, x, y, z (all positive integers) where exponents u, v, x, y, z are greater than 2, of the following Beal Equation

$A^u + B^v + C^x + D^y = E^z$

Given z = 15, u,v,x,y > 2; and

$\{gcf(A,B,C,D,E) = P\}$

*) Find the values of A, B, C, D, E, u, v, x, y, z (all positive integers) where exponents u, v, x, y, z are greater than 2, of the following Beal Equation

$$A^u + B^v + C^x + D^y = E^z$$

Given u = 5, v,x,y,z > 2; and

$$\{gcf(A,B,C,D,E) = P\}$$

*) Find the values of A, B, C, D, E, u, v, x, y, z (all positive integers) where exponents u, v, x, y, z are greater than 2, of the following Beal Equation

$$A^u + B^v + C^x + D^y = E^z$$

Given z = 14; and

$$\{gcf(A,B,C,D,E) = P\}$$

*) Find the values of A, B, C, D, E, u, v, x, y, z (all positive integers) where exponents u, v, x, y, z are greater than 2, of the following Beal Equation

$$A^u + B^v + C^x + D^y = E^z$$

Given z = 8 ; and

$$\{gcf(A,B,C,D,E) = P\}$$

# EXPANSION OF THE BEAL CONJECTURE TO FORM

$$A^t + B^u + C^v + D^x + E^y = F^z$$

We have completed proof of the Beal Conjecture, so the hypothesis becomes Beal's Theorem. Now we use the Beal's Theorem in the form of the following equation:

$$A^r + B^s + C^t + D^u + E^v + F^x + G^y = H^z$$

Values of A, B, C, D, E, F, G, H, (positive integers), which must have a common factor of a prime number, and the exponents r, s, t, u, v, x, y, z are greater than 2

*) Find the values of A, B, C, D, E, F, G, H and r, s, t, u, v, x, y, z (all positive integers) where exponents r,

s, t, u, v, x, y, z are greater than 2, of the following Beal Equation

$$A^r + B^s + C^t + D^u + E^v + F^x + G^y = H^z$$

Know that r, s, t, u, v, x, y, z > 2;

And {gcf(A,B,C,D,E,F,G,H) = P}

Prove

$$A^r + B^s + C^t + D^u + E^v + F^x + G^y = H^z$$

Applying the new formula for Beal Conjecture

$$(A,B,C) \in \mathbf{Z}$$

$$(x,y,z) \in \mathbf{Z} > 2$$

$$C \equiv \sqrt[z]{A^{modz}} \sqrt[z]{(\Delta - 1) + modZ^{|y-x|}}$$

$\Delta$: *left term's Equation*

$$\rightarrow gcf(A,B,C) \geq 2$$

We have

$$792280491775295473025328100000^6 + 281474775384100^{12} + 16777210^{24} + 4722361416321876361000^8 + 281474775384100^{12} + 4722361416321876361000^8 + 16777210^{25} = 33554420^{24}$$

Try again

Left equation

$79228049177529547302532810000^{6} + 281474775384100^{12} + 16777210^{24} + 472236141632187636100 0^{8} + 281474775384100^{12} + 4722361416321876361000^{8} + 16777210^{25} = 4.149479953449657363922608853768 8e+180$

Right equation

$33554420^{24} = 4.1494799534496573639226088537688e+180$

Solution

| | |
|---|---|
| A = 79228049177529547302532810000 | r = 6 |
| B = 281474775384100 | s = 12 |
| C = 16777210 | t = 24 |
| D = 4722361416321876361000 | u = 8 |
| E = 281474775384100 | v = 12 |
| F = 4722361416321876361000 | x = 8 |
| G = 16777210 | y = 25 |
| H = 33554420 | z = 24 |

# EXPANSION OF THE BEAL CONJECTURE TO FORM

## $A^p + B^q + C^r + D^s + E^t + F^u + G^v + H^x + I^y = J^z$

We apply the new formula

$$(A, B, C) \in \boldsymbol{Z}$$

$$(x, y, z) \in \boldsymbol{Z} > 2$$

$$C \equiv \sqrt[z]{A^{modz}}\ \sqrt[z]{(\Delta - 1) + modZ^{|y-x|}}$$

Δ: *left term's Equation*

$$\rightarrow gcf(A, B, C) \geq 2$$

*) Find the values of A, B, C, D, E, F, G, H, I, J and the exponents p, q, r, s, t, u, v, x, y, z, using the Andy

Beal conjecture on the condition of the following Diophantine equation

$$A^p + B^q + C^r + D^s + E^t + F^u + G^v + H^x + I^y = J^z$$

Prove

Apply the new formula

$$(A,B,C) \in \boldsymbol{Z}$$

$$(x,y,z) \in \boldsymbol{Z} > 2$$

$$C \equiv \sqrt[z]{A^{modz}} \; \sqrt[z]{(\Delta - 1) + modZ^{|y-x|}}$$

$\Delta$: *left term's Equation*

$$\rightarrow gcf(A,B,C) \geq 2$$

We have many solutions

Ex:

$$79228011398644972121594269696^6 + 281474708275264^{12} + 4722359727473425382912^8 + 4722359727473425382912^8 + 4722359727473425382912^8 + 4722359727473425382912^8 + 281474708275264^{12} + 281474708275264^{12} + 16777208^{25} = 33554416^{24}$$

Left side

$79228011398644972121594269696^{6} + 281474708275264^{12} + 4722359727473425382912^{8} + 4722359727473425382912^{8} + 4722359727473425382912^{8} + 4722359727473425382912^{8} + 281474708275264^{12} + 281474708275264^{12} + 16777208^{25}$
$= 4.149468081704312619176682122494e+180$

Right side

$33554416^{24} = 4.149468081704312619176682122494e+180$

# OTHER METHODS

We use the new General method for the Beal conjecture.

$$(A, B, C) \in Z$$

$$(x, y, z) \in Z > 2$$

$$C \equiv \sqrt[z]{A^{modz}} \sqrt[z]{(\Delta - 1) + modA^{|y-x|}}$$

$\Delta$: *left term's Equation*

$$\rightarrow gcf(A, B, C) \geq 2$$

# METHOD II FOR BEAL CONJECTURE

$$A^x + B^y = C^z$$

The Diophantine equation has form *BEAL'S CONJECTURE,*

$$A^x + B^y = C^z$$

And

$$x, y, z > 2,$$

Often three exponents of the equation are equal, (i.e. x, x, x, or 3, 3, 3; 4, 4, 4; or 5, 5, 5; . . .) or not equal (i. e. x, y, z, or 3, 4, 5; 7, 5, 4; . . .), but this problem we only know ($x, y, z > 2$).

It's hard, but we also have many methods for solution of the values of A, B, C, and x, y, z where they are all positive integers, and the exponents are all greater than 2,

### *Popular Method*

Choose any values of a, b, c for:

$$a + b = c \qquad (1)$$

Multiply both sides of equation (1) above for the number of "$a^i\, b^j\, c^k$" then equation (1) is not changed

$$a^{i+1} \cdot b^j \cdot c^k + a^i \cdot b^{j+1} \cdot c^k = a^i \cdot b^j \cdot c^{k+1}$$

,

Conditions

gcf(i+1, j, k) = x
gcf(i, j+1, k) = y
gcf(i, j, k+1) = z

This equation is rewritten in the form of Beal Conjecture

$$(a^{(i+1)'} \cdot b^{j'} \cdot c^{k'})^x + (a^{i'} \cdot b^{(j+1)'} \cdot c^{k'})^y = (a^{i'} \cdot b^{j'} \cdot c^{(k+1)'})^z$$

We have

$$A^x = (a^{(i+1)'} \cdot b^{j'} \cdot c^{k'})^x$$
$$B^y = (a^{i'} \cdot b^{(j+1)'} \cdot c^{k'})^y$$
$$C^z = (a^{i'} \cdot b^{j'} \cdot c^{(k+1)'})^z$$

Back to the original form of the Beal Conjecture

$$A^x + B^y = C^z$$

# METHOD III FOR BEAL CONJECTURE

$$A^x + B^y = C^z$$

The popular method is difficult, but the following method is very easy for students.

$$\sum_{k\geq 2}^{n} a = a^k$$

$$\sum_{k=1}^{n} a = b^k$$

$$\sum_{k\geq 2}^{n} a = a.b^k$$

Many thousands of years ago, the Ancient Greeks and Babylonians knew these formulas

Example:

These Formulas

$$\sum_{k=0}^{n}\binom{n}{k}=2^{n}$$

$$\sum_{and\ k=0}^{n} k^{2}\binom{n}{k}=(n+n^{2})2^{n-2}$$

Now I rewrite these Formulas for the Beal Conjecture . . .

$$a^{m}\sum_{k\geq 2}^{n} a=a^{m+k}$$

$$a^{m}\sum_{k\geq 2}^{n} a=b^{m+k}$$

$$a^{m}b^{q}\sum_{k\geq 2}^{n} a=a^{m+1}b^{q+k}$$

If the formula is too simple, it is difficult for the equations of the form Beal Conjecture

$$A^{x}+B^{y}=C^{z}$$

So I added little value to formula

For example:
This formula

$$\sum_{k\geq 2}^{n} a = a^{k}$$

Multiply both sides for $a^m$

$$a^{m}\sum_{k\geq 2}^{n} a = a^{m+k}$$

Similarly with

$$\sum_{k\geq 2}^{n} a = b^{k}$$

Multiply both sides for $b^m$
We have

$$b^{m}\sum_{k\geq 2}^{n} a = b^{m+k}$$

Similarly

$$\sum_{k\geq 2}^{n} a = a.\, b^{k}$$

Multiply both sides for $a^m b^q$

$$a^m b^q \sum_{k \geq 2}^{n} a = a^{m+1} b^{q+k}$$

. . . .

The formulas are the Ancient methods, so we do not need to prove them

$$a^m \sum_{k \geq 2}^{n} a = a^{m+k}$$

$$b^m \sum_{k \geq 2}^{n} a = b^{m+k}$$

$$a^m b^q \sum_{k \geq 2}^{n} a = a^{m+1} b^{q+k}$$

# BEAL EQUATION

$$A^x + B^y = C^z$$

*) Find the values of A, B, C (positive integers) which must have a common prime factor and the exponents x, y, z are all greater than 2

$$A^x + B^y = C^z$$

Prove

We have a Diophantine Equation

$$A^x + B^y = C^z$$

***Method I***

For any values of a and b we have

$$a + b = c$$

Ex:

$$1 + 3 = 4 = 2^2 \quad (2)$$

Multiply both side of the equation (2) for "$2^{30} \cdot 3^{24}$"

$$(2^{30} \cdot 3^{24}) + (2^{30} \cdot 3^{25}) = (2^{32} \cdot 3^{24})$$

Rewrite to form Beal equation

$$(2^5 \cdot 3^4)^6 + (2^6 \cdot 3^5)^5 = (2^4 \cdot 3^3)^8$$

$$A^x + B^y = C^z$$

We have

$A = (2^5 \cdot 3^4) = 2592,$ $x = 6$
$B = (2^6 \cdot 3^5) = 15552,$ $y = 5$
$C = (2^4 \cdot 3^3)^8 = 432,$ $z = 8$

To make sure, we must try again

$$2592^6 + 15552^5 = 432^8$$

Left equation

$$2592^6 + 15552^5 = 1213025622610333925376$$

Right equation

$$432^8 = 1213025622610333925376$$

Solution:

$A = 2592,$ $\quad x = 6$
$B = 15552,$ $\quad y = 5$
$C = 432,$ $\quad z = 8$

. . . .

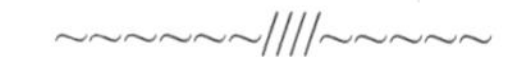

*) Find the values of A, B, C (positive integers) which must have a common prime factor and the exponents x, y, z are all greater than 2

$$A^x + B^y = C^z$$

~~~~ Prove ~~~~

Similarly, for any value of a and b

We have

$$a + b = c$$

Ex:

$a = 1$, $b = 7$

$$1 + 7 = 8 = 2^3$$

Multiply both side of the equation for $2^{24} \cdot 7^{15}$
~~~~

$2^{24 \cdot}7^{15} + 2^{24 \cdot}7^{16} = 2^{27 \cdot}7^{15}$

$(2^{8 \cdot}7^{5})^{3} + (2^{3 \cdot}7^{2})^{8} = (2^{9 \cdot}7^{5})^{3}$

$A^{x} + B^{y} = C^{z}$

$4302592^{3} + 392^{8} = 8605184^{3}$

To make sure, we must try again

Left equation

$4302592^{3} + 392^{8} = 637206919404798869504$

Right equation

$8605184^{3} = 637206919404798869504$

Solution:

A = 4302592, x = 3
B = 392, y = 8
C = 8605184, z = 3

. . . .

~~~~~~/////~~~~~
~~~~~~

# BEAL EQUATION

$$A^x + B^y = C^z$$

*) Find the values of A, B, C (positive integers) which must have a common prime factor and the exponents x, y, z are all greater than 2

$$A^x + B^y = C^z$$

Prove

We have a Diophantine Equation

$$A^x + B^y = C^z$$

***Method II***

If a = 2 and n = 2
We use

$$\sum_{k=2}^{n} a = a^2$$

Multiply both side of the equation for $a^m$

$$a^m \sum_{k \geq 2}^{n=2} a = a^{m+2}$$

If we choose: $a^m = 2^{14}$

$$2^{14} \sum_{k \geq 2}^{n=2} 2 = 2^{14+2}$$

$$\sum_{k \geq 2}^{n=2} 21\text{^}15 = 2^{16}$$

Or

$$2^{15} + 2^{15} = 2^{16}$$

Rewrite to form the Beal Conjecture

$$(2^5)^3 + (2^3)^5 = (2^4)^4$$

$$32^3 + 8^5 = 16^4$$

Solution

A = 32, x = 3
B = 8, y = 5
C = 16, z = 4

~~~~~/////~~~~~
~~~~~

*) Find the values of A, B, C (positive integers) which must have a common prime factor and the exponents x, y, z are all greater than 2

$$A^x + B^y = C^z$$

~~~~ Prove ~~~~

Applying this formula for the Beal's conjecture

$$\sum_{k=1}^{n} a = a \cdot b^k$$

With n = 2 and a = 3

$$\sum_{k=1}^{n=2} 3 = 3 \cdot 2^1$$

Multiply both side of the equation for $a^m b^q$

$$a^m \cdot b^q \sum_{k=1}^{n=2} a = a^m \cdot b^{q+1}$$

We choose

$$a^m \cdot b^q = 3^{29} \cdot 2^{24}$$

$$3^{29} \cdot 2^{24} \sum_{k=1}^{n=2} 3 = 3^{30} \cdot 2^{25}$$
~~~~

$$\sum_{k=1}^{n=2} 3^{30} \cdot 2^{24} = 3^{30} \cdot 2^{25}$$

We can rewrite

$$(2^{8} 3^{10})^{3} + (2^{4} 3^{5})^{6} = (2^{5} 3^{6})^{5}$$

$$15116544^{3} + 3888^{6} = 23328^{5}$$

To make sure, we must try again

Left equation

$$15116544^{3} + 3888^{6} = 6908559991272917434368$$

Right equation

$$23328^{5} = 6908559991272917434368$$

Solution

A = 15116544, x = 3
B = 3888, y = 6
C = 23328, z = 5

~~~~~/////~~~~~

*) Find the values of A, B, C, which must have a common prime factor and the exponent x, y, z greater than 2
~~~~~

$$A^x + B^y = C^z$$

~~~~ Prove ~~~~

Applying this formula for the Beal's conjecture

$$\sum_{k=1}^{n} a = a \cdot b^k$$

n = 2, a = 13

$$\sum_{k=1}^{n=2} 13 = 13 \cdot b^1$$

Multiply both side of the equation for $a^m b^q$

$$a^m \cdot b^q \sum_{k=1}^{n=2} a = a^m \cdot b^{q+1}$$

If $a^m \cdot b^q = 13^{29} \cdot 2^{24}$

We have

$$13^{29} \cdot 2^{24} \sum_{k=1}^{n=2} 13 = 13^{30} \cdot 2^{25}$$

or
~~~~

$$\sum_{k=1}^{n=2} 13^{30} \cdot 2^{24} = 13^{30} \cdot 2^{25}$$

We rewrite to form Beal conjecture

$$(2^{8}.13^{10})^{3} + (2^{4}.13^{5})^{6} = (2^{5}.13^{6})^{5}$$

$$35291773913344^{3} + 5940688^{6} = 154457888^{5}$$

To make sure, we must try again

Left equation

$35291773913344^{3} + 5940688^{6}$
$= 8.7912465665148309976831907175499e+40$

Right equation

$154457888^{5} = 8.7912465665148309976831907175499e+40$

Solution

A = 35291773913344, x = 3
B = 5940688, y = 6
C = 154457888, z = 5

*) Find the values of A, B, C, which must have a common prime factor and the exponents x, y, z greater than 2

$$A^x + B^y = C^z$$

~~~~ Prove ~~~~

Similarly above

$$\sum_{k=1}^{n} a = a \cdot b^k$$

With n = 2 and a = 17

$$\sum_{k=1}^{2} 17 = 17 \cdot 2^1$$

Multiply both side of the equation for "$a^m \cdot b^q$"

$$\sum_{k=1}^{n} a = a \cdot b^k$$

$$a^m \cdot b^q \sum_{k=1}^{n=2} a = a^m \cdot b^{q+1}$$

We choose $a^m b^q = 17^{71} 2^8$

$$17^{71} \cdot 2^8 \sum_{k=1}^{n=2} 17 = 17^{72} \cdot 2^9$$
~~~~

$$\sum_{k=1}^{n=2} 17^{72} \cdot 2^{8} = 17^{72} \cdot 2^{9}$$

$$2^{8} \cdot 17^{72} + 2^{8} \cdot 17^{72} = 2^{9} \cdot 17^{72}$$

We rewrite to form Beal conjecture

$$(2 \cdot 17^{9})^{8} + (2^{2} \cdot 17^{18})^{4} = (2 \cdot 17^{8})^{9}$$

$$237175752994^{8} + 56252337808270899964036^{4} = 13951514882^{9}$$

To make sure, we must try again

Left equation

$$237175752994^{8} + 56252337808270899964036^{4} = 2.00259118524976279949936743945 8e+91$$

Right equation

$$13951514882^{9} = 2.002591185249762799499367439458e+91$$

Solution

A = 237175752994, x = 8
B = 56252337808270899964036, y = 4
C = 13951514882, z = 9

*) Find the values of A, B, C, which must have a common prime factor and the exponents x, y, z greater than 2

$$A^x + B^y = C^z$$

~~~~ Prove ~~~~

Similarly above

$$\sum_{k=1}^{n} a = a \cdot b^k$$

With n = 2 and a = 23

$$\sum_{k=1}^{2} 23 = 23 \cdot 2^1$$

Multiply both side of the equation for $a^m \cdot b^q$

$$\sum_{k=1}^{n} a = a \cdot b^k$$

$$a^m \cdot b^q \sum_{k=1}^{n=2} a = a^m \cdot b^{q+1}$$
~~~~

We choose $a^m \cdot b^q = 23^{41} 2^6$

$$23^{41} \cdot 2^6 \sum_{k=1}^{n=2} 23 = 23^{42} \cdot 2^7$$

$$\sum_{k=1}^{n=2} 23^{42} \cdot 2^6 = 23^{42} \cdot 2^7$$

We rewrite to form Beal conjecture

$$2^6 \cdot 23^{42} + 2^6 \cdot 23^{42} = 2^7 \cdot 23^{42}$$

Or

$$(2 \cdot 23^7)^6 + (2^2 \cdot 23^{14})^3 = (2 \cdot 23^6)^7$$

$6809650894^6 + 463713452981549992363^3 = 296071778^7$

To make sure, we must try again

Left equation

$6809650894^6 + 463713452981549992363^3$
$= 1.9942476198363392432719023681265e+59$

Right equation

$296071778^7 = 1.9942476198363392432719023681265e+59$

Solution

$A = 6809650894,$ $\qquad x = 6$
$B = 463713452981549992 36,$ $\qquad y = 3$
$C = 296071778,$ $\qquad z = 7$

We apply this method

$$a^m b^q \sum_{k \geq 2}^{n} a = a^{m+1} b^{q+k}$$

For the Beal Conjecture and following problems

*) Find the values of A, B, C, D, E, F, G, H, I, J, p, q, r, s,t, u, v, x, y, z which are all positive integers and exponents p, q, r, s,t, u, v, x, y, z are greater than 2, of the following Beal Equation

$$A^p + B^q + C^r + D^s + E^t + F^u + G^v + H^x + I^y = J^z$$

Know that p,q,r,s,t, u,v,x,y,z > 2; and

$$\{gcf(A,B,C,D,E,F,G,H,I,J) = P\}$$

Diophantine Equation:

$$A^p + B^q + C^r + D^s + E^t + F^u + G^v + H^x + I^y = J^z$$

There are 20 unknowns.
We apply one of three formulas,

$$a^m b^q \sum_{k\geq 2}^{n} a = a^{m+1} b^{q+k}$$

We have the following result

$$387420489^8 + 81^{36} + 6561^{18} + 43046721^9 + 531441^{12} + 282429536481^6 + 729^{24} + 150094635296999121^4 + 797664430768725098633613^3 = 3^{146}$$

Try again

Left equation

$$387420489^8 + 81^{36} + 6561^{18} + 43046721^9 + 531441^{12} + 282429536481^6 + 729^{24} + 150094635296999121^4 + 797664430768725098633613^3$$
$$= 4.567759074507740406477787437675 3e+69$$

Right equation

$$3^{146} = 4.5677590745077404064777874376753e+69$$

Solution

A = 387420489 p = 8
B = 81 q = 36
C = 6561 r = 18
D = 43046721 s = 9
E = 531441 t = 12

| | |
|---|---|
| F = 282429536481 | u = 6 |
| G = 729 | v = 24 |
| H = 150094635296999121 | x = 4 |
| I = 79766443076872509863361 | y = 3 |
| J = 3 | z = 146 |

…

# EXERCISE

*) find the positive integers of A, B, C, which must have a common factor for prime number and the exponent x, y, z are greater than 2

$$A^x + B^y = C^z$$

Give: $x = 3$, $y = 5$, $z = 4$

*) Find the positive integers of A, B, C, which must have a common factor of a prime number and the exponents x, y, z are greater than 2

$$A^x + B^y = C^z$$

Given: $x = 4$, $y = 7$, $z = 6$

*) Find the positive integers of A, B, C, which must have a common factor of a prime number and the exponents x, y, z are greater than 2

$$A^x + B^y = C^z$$

Given: $x = 5$, $y = 5$, $z = 6$

*) Find the positive integers of A, B, C, which must have a common factor of a prime number and the exponents x, y, z are greater than 2

$$A^x + B^y = C^z$$

Given: $x = 6$, $y = 5$, $z = 7$

*) Find the positive integers of A, B, C, which must have a common factor of a prime number and the exponents x, y, z are greater than 2

$$A^x + B^y = C^z$$

Given: $x = 7$, $y = 4$, $z = 5$

*) Find the positive integers of A, B, C, which must have a common factor of a prime number and the exponents x, y, z are greater than 2

$$A^x + B^y = C^z$$

Given: $x = 4$, $y = 6$, $z = 7$

*) Find the positive integers of A, B, C, which must have a common factor of a prime number and the exponents x, y, z are greater than 2

$$A^x + B^y = C^z$$

Given: $x = 7$, $y = 5$, $z = 8$

*) Find the positive integers of A, B, C, which must have a common factor of a prime number and the exponents x, y, z are greater than 2

$$A^x + B^y = C^z$$

Given: $x = 8, y = 4, z = 7$

*) Find the positive integers of A, B, C, which must have a common factor of a prime number and the exponents x, y, z are greater than 2

$$A^x + B^y = C^z$$

Given: $x = 9, y = 5, z = 8$

*) Find the positive integers of A, B, C, which must have a common factor of a prime number and the exponents x, y, z are greater than 2

$$A^x + B^y = C^z$$

Given: $x = 8, y = 5, z = 7$

*) Find the positive integers of A, B, C, which must have a common factor of a prime number and the exponents x, y, z are greater than 2

$$A^x + B^y = C^z$$

Given: $x = 9, y = 4, z = 7$

*) Find the positive integers of A, B, C, which must have a common factor of a prime number and the exponents x, y, z are greater than 2

$$A^x + B^y = C^z$$

Given: x = 10, y = 5, z = 7

*) Find the positive integers of A, B, C, which must have a common factor of a prime number and the exponents x, y, z are greater than 2

$$A^x + B^y = C^z$$

Given: x = 8, y = 4, z = 10

# EXPANSION OF THE BEAL CONJECTURE TO FORM

$$A^v + B^x + C^y = D^z$$

We have completed proof of Beal Conjecture, so the hypothesis becomes Beal's Theorem. Now we use Beal's Theorem in the form of the following equation:

$$A^v + B^x + C^y = D^z$$

Values of A, B, C, D are positive integers,(not zero), which must have a common factor of a prime number and the exponents x, y, z are greater than 2

*) Find the values of A, B, C, D (positive integers), which must have a common factor of a prime number and the exponents v, x, y, z are greater than 2 of in following equation:

$$A^v + B^x + C^y = D^z$$

~~~~~Prove~~~~~

Apply the following formula

$$\sum_{k\geq 2}^{n} a = a^k$$

With n = 3 and a = 3:

$$\sum_{k=2}^{n=3} 3 = 3\cdot 3 = 3^2$$

Multiply 2 sides for the $a^m$

$$a^m \sum_{k=2}^{n=3} a = a^{m+2}$$

Replace $a^m = 3^{11}$

$$3^{11} \sum_{k=2}^{n=3} 3 = 3^{13}$$
~~~~~

Rewritten as

$$3 + 3 + 3 = 3^2 \qquad (1b)$$

Multiply both sides of the equation (1b) by $3^{11}$

$$3^{12} + 3^{12} + 3^{12} = 3^{13}$$

Rewritten as the Beal's equation

$$(3^2)^6 + (3^3)^4 + (3^4)^3 = 3^{13}$$

Then $\quad 9^6 + 27^4 + 81^3 = 3^{13}$

Solution

A = 9, v = 6
B = 27, x = 4
C = 81, y = 3
D = 3, z = 13

*) Find the values of A, B, C, D (positive integers), which must have a common factor of a prime number and the exponents v, x, y, z are greater than 2 in the following equation:

$$A^v + B^x + C^y = D^z$$

~~~~~Prove~~~~~

Apply the following formula

$$\sum_{k \geq 2}^{n} a = b \cdot a^{k}$$

With n = 3 and a =

$$\sum_{k=1}^{n=3} 5 = 3 \cdot 5 = 3 \cdot 5^{1}$$

Multiply 2 sides for $a^{m} \cdot b^{n}$

$$\sum_{k=1}^{n=3} a = b \cdot a^{k}$$

$$a^{m} \cdot b^{q} \sum_{k=1}^{n=3} a = a^{m} \cdot b^{q+1}$$

Given $a^{m} \cdot b^{q} = 5^{59} \cdot 3^{24}$

$$5^{59} \cdot 3^{24} \sum_{k=1}^{n=3} 5 = 5^{60} \cdot 3^{25}$$

Write the following simple equation

$$5 + 5 + 5 = 3 \cdot 5 \qquad (2b)$$
~~~~~

Multiply both sides of the equation (2b) by $3^{24}.5^{59}$

We have

$$3^{24}.5^{60} + 3^{24}.5^{60} + 3^{24}.5^{60} = 3^{25}.5^{60}$$

Rewritten as the Beal equation

$$(3^{2}.5^{5})^{12} + (3^{4}.5^{10})^{6} + (3^{6}.5^{15})^{4} = (3^{5}.5^{12})^{5}$$

$$28125^{12} + 791015625^{4} + 22247314453125^{4} = 59326171875^{5}$$

Try again

The left side of the equation

$$28125^{12} + 791015625^{4} + 22247314453125^{4}$$
$$= 7.3490572086425814957166835483804e+53$$

The right side of the equation

$$59326171875^{5} = 7.3490572086425814957166835483804e+53$$

Solution

A = 28125, v = 12
B = 791015625, x = 6
C = 22247314453125, y = 4
D = 59326171875, z = 5

*) Find the values of A, B, C, D (positive integers), which must have a common factor of a prime number and the exponents v, x, y, z are greater than 2 in the following equation:

$$A^v + B^x + C^y = D^z$$

~~~~~Prove~~~~~

Apply the following formula

$$\sum_{k\geq 2}^{n} a = b \cdot a^k$$

n = 3 and a = 7
We have

$$\sum_{k=1}^{n=3} 7 = 3 \cdot 7 = 3 \cdot 7^1$$

Multiply 2 sides for $a^m \cdot b^n$

$$\sum_{k=1}^{n=3} a = b \cdot a^k$$

$$a^m \cdot b^q \sum_{k=1}^{n=3} a = a^m \cdot b^{q+1}$$
~~~~~

Given $a^m \cdot b^q = 7^{155} \cdot 3^{12}$

$$3^{12} \cdot 7^{155} \sum_{k=1}^{n=3} 7 = 7^{156} \cdot 3^{13}$$

Write the following simple equation

$$7 + 7 + 7 = 3 \cdot 7 \text{ (3b)}$$

Multiply both sides of the equation (3b) by $3^{12} \cdot 7^{155}$

$$3^{12} \cdot 7^{156} + 3^{12} \cdot 7^{156} + 3^{12} \cdot 7^{156} = 3^{13} \cdot 7^{156}$$

Rewriting the above equation in the form of Beal

$$(3 \cdot 7^{13})^{12} + (3^2 \cdot 7^{26})^6 + (3^3 \cdot 7^{39})^4 = (3 \cdot 7^{12})^{13}$$

$290667031221^{12} + 844873230388297887508416^{6} + (2.4557679363506250802145535537007e+34)^4 = 41523861603^{13}$

Try again

The left side of the equation

$290667031221^{12} + 844873230388297887508416^{6} + (2.4557679363506250802145535537007e+34)^4 = 1.0911150686937738321656834008995e+138$

The right side of the equation

$41523861603^{13} = 1.0911150686937738321656834008995e+138$

The positive integers of the values of A, B, C, D, and exponent greater than 2, which correct to the Beal Conjecture

Solution

A = 28125, v = 12
B = 8448732303882978875 0841, x = 6
C = 2.45576793635062508021455355537007e+34, y = 4
D = 41523861603, z = 13

~~~~~/////~~~~~
~~~~~

# EXERCISE

*) Find the positive integers A, B, C, D which must have a common factor of a prime number and the exponents v, x, y, z are greater than 2

$$A^v + B^x + C^y = D^z$$

Given: $v = 3$, $y = 4$

*) Find the positive integers A, B, C, D which must have a common factor of a prime number and the exponents v, x, y, z are greater than 2 of the following equation:

$$A^v + B^x + C^y = D^z$$

Given: $v = 5$, $y = 4$

*) Find the positive integers A, B, C, D which must have a common factor of a prime number and the exponents v, x, y, z are greater than 2 of the following equation:

$$A^v + B^x + C^y = D^z$$

Given: $v = 4$, $y = 5$

*) Find the positive integers A, B, C, D which must have a common factor of a prime number and the exponent v, x, y, z are greater than 2 of the following equation:

$$A^v + B^x + C^y = D^z$$

Given: $v = 3$, $y = 7$

*) Find the positive integers A, B, C, D which must have a common factor of a prime number and the exponent v, x, y, z are greater than 2 of the following equation:

$$A^v + B^x + C^y = D^z$$

Given: $v = 4$, $y = 6$

*) Find the positive integers A, B, C, D which must have a common factor of a prime number and the exponents v, x, y, z are greater than 2 of the following equation:

$$A^v + B^x + C^y = D^z$$

Given: $v = 3$, $y = 4$, $z = 7$

*) Find the positive integers A, B, C, D which must have a common factor of a prime number and the exponents v, x, y, z are greater than 2 of the following equation:

$$A^v + B^x + C^y = D^z$$

Given: $v = 5$, $y = 4$, $z = 6$

*) Find the positive integers A, B, C, D which must have a common factor of a prime number and the exponents v, x, y, z are greater than 2 of the following equation:

$$A^v + B^x + C^y = D^z$$

Given: v = 6, y = 4, y = 7

*) Find the positive integers A, B, C, D which must have a common factor of a prime number and the exponents v, x, y, z are greater than 2 of the following equation:

$$A^v + B^x + C^y = D^z$$

Given: v = 7, y = 4, z = 3

*) Find the positive integers A, B, C, D which must have a common factor of a prime number and the exponent v, x, y, z are greater than 2 of the following equation:

$$A^v + B^x + C^y = D^z$$

Given: v = 7, y = 6

*) Find the positive integers A, B, C, D which must have a common factor of a prime number and the exponents v, x, y, z are greater than 2 of the following equation:

$$A^v + B^x + C^y = D^z$$

Given: v = 8, y = 5

*) Find the positive integers A, B, C, D which must have a common factor of a prime number and the exponents v, x, y, z are greater than 2 of the following equation:

$$A^v + B^x + C^y = D^z$$

Given: $v = 8, y = 4$

*) Find the positive integers A, B, C, D which must have a common factor of a prime number and the exponents v, x, y, z are greater than 2 of the following equation:

$$A^v + B^x + C^y = D^z$$

Given: $v = 5, y = 4, z = 8$

*) Find the positive integers A, B, C, D which must have a common factor of a prime number and the exponents v, x, y, z are greater than 2 of the following equation:

$$A^v + B^x + C^y = D^z$$

Given: $v = 3, x = 4, y = 5$

*) Find the positive integers A, B, C, D which must have a common factor of a prime number and the exponents v, x, y, z are greater than 2 of the following equation:

$$A^v + B^x + C^y = D^z$$

Given: $v = 4, x = 7, y = 5$

*) Find the positive integers A, B, C, D which must have a common factor of a prime number and the exponents v, x, y, z are greater than 2 of the following equation:

$$A^v + B^x + C^y = D^z$$

Given: v = 7, x = 5, y = 7

*) Find the positive integers A, B, C, D which must have a common factor of a prime number and the exponents v, x, y, z are greater than 2 of the following equation:

$$A^v + B^x + C^y = D^z$$

Given: v = 6, x = 4, y = 5, z = 4

*) Find the positive integers A, B, C, D which must have a common factor of a prime number and the exponent v, x, y, z are greater than 2 of the following equation:

$$A^v + B^x + C^y = D^z$$

Given: v = 4, x = 4, y = 5, z = 7

*) Find the positive integers A, B, C, D which must have a common factor of a prime number and the exponent v, x, y, z are greater than 2 of the following equation:

$$A^v + B^x + C^y = D^z$$

Given: v = 7, x = 6, y = 5, z = 8

*) Find the positive integers A, B, C, D which must have a common factor of a prime number and the exponents v, x, y, z are greater than 2 of the following equation:

$$A^v + B^x + C^y = D^z$$

Given: v = 6, x = 4, y = 5, z = 7

*) Find the positive integers A, B, C, D which must have a common factor of a prime number and the exponents v, x, y, z are greater than 2 of the following equation:

$$A^v + B^x + C^y = D^z$$

Given: v = 7, x = 4, y = 5, z = 6

*) Find the positive integers A, B, C, D, which must have a common factor of a prime number and the exponents v, x, y, z are greater than 2 of the following equation:

$$A^v + B^x + C^y = D^z$$

Given: v = 3, x = 4, y = 5, z = 6

# EXPANSION OF THE BEAL CONJECTURE TO FORM

$$A^u + B^v + C^x + D^y = E^z$$

We have completed proof of Beal Conjecture, so the hypothesis becomes Beal's Theorem. Now we use Beal's Theorem in the form of the following equation:

$$A^u + B^v + C^x + D^y = E^z$$

Values of A, B, C, D, E (positive integers), which must have a common factor of a prime number, and the exponents u, v, x, y, z are greater than 2

*) Find the positive integers A, B, C, D, E, which must have a common factor of a prime number and the exponents u, v, x, y, z are greater than 2 of the following equation:

$$A^u + B^v + C^x + D^y = E^z$$

~~~~~Prove~~~~~

Apply the following formula

$$\sum_{k=2}^{n=4} a = b^2$$

n = 4 and a = 1
We have

$$\sum_{k=2}^{n=4} 1 = 2^2$$

Multiply both sides by $b^m$

$$b^m \sum_{k=2}^{n=4} a = b^{m+2}$$

Choose: $b^m = 2^{36}$

$$2^{36} \sum_{k=2}^{n=4} 1 = 2^{38}$$
~~~~~

Writing simple form to the equation as follows

$$1 + 1 + 1 + 1 = 2^2 \qquad (1c)$$

Multiply both sides of the equation (1c) by $2^{36}$

$$2^{36} + 2^{36} + 2^{36} + 2^{36} = 2^{36+2}$$

Rewriting the above equation in the form of Beal

$$16^9 + 4^{18} + 8^{12} + 64^6 = 4^{19}$$

The left side of the equation

$16^9 + 4^{18} + 8^{12} + 64^6 = 274877906944$

The right side of the equation

$$4^{19} = 274877906944$$

Solution

A = 16, u = 9
B = 4, v = 18
C = 8, x = 12
D = 64 y = 6
E = 4 z = 19

*) Find the positive integers A, B, C, D, E, which must have a common factor of a prime number and the exponents u, v, x, y, z are greater than 2 of the following equation:

$$A^u + B^v + C^x + D^y = E^z$$

~~~~~Prove~~~~~

Apply the following formula

$$\sum_{k=2}^{n=4} a = a^2$$

Put a = 4, n = 4

$$\sum_{k=2}^{n=4} 4 = 4^2$$

Multiply both sides by $a^m$

$$a^m \sum_{k=2}^{n=4} a = a^{m+2}$$

Choose $a^m = 4^{23}$

$$4^{23} \sum_{k=2}^{n=4} 4 = 4^{25}$$
~~~~~

Writing simple form to the equation as follows

$$4 + 4 + 4 + 4 = 4^2 \qquad (2c)$$

Multiply both sides of the equation (2c) by $4^{23}$

$$4^{24} + 4^{24} + 4^{24} + 4^{24} = 4^{25}$$

Rewriting the above equation in the form of Beal

$$16^{12} + 256^6 + 2^{48} + 64^8 = 3125^5$$

Solution

A = 16, u = 12
B = 256, v = 6
C = 2, x = 48
D = 64, y = 8
E = 1024, z = 5

~~~~~/////~~~~~

*) Find the positive integers A, B, C, D, E, which must have a common factor of a prime number and the exponents u, v, x, y, z are greater than 2 of the following equation:

$$A^u + B^v + C^x + D^y = E^z$$
~~~~~

~~~~~Prove~~~~~

Apply the following formula (a = 2 and n =4)

$$\sum_{k\geq 2}^{n} a = a \cdot b^{k}$$

Rewrite a = 2 and n =4

$$\sum_{k\geq 2}^{n=4} 2 = 4 \cdot 2 = 2^{3}$$

Multiply both sides by $2^{23}$

$$2^{23}\sum_{k=2}^{n=4} 2 = 2^{26}$$

Writing simple form to the equation as follows

$$2 + 2 + 2 + 2 = 2^{3} \qquad (3c)$$

Multiply both sides of the equation (3c) by $2^{23}$

$$2^{24} + 2^{24} + 2^{24} + 2^{24} = 2^{26}$$

Rewriting the above equation in the form of Beal

$$(2^{8})^{3} + (2^{4})^{6} + (2^{6})^{4} + (2^{3})^{8} = (2^{2})^{13}$$

$$256^{3} + 16^{6} + 64^{4} + 8^{8} = 4^{13}$$
~~~~~

Solution

| | |
|---|---|
| A = 256, | u = 3 |
| B = 16, | v = 6 |
| C = 64, | x = 4 |
| D = 8, | y = 8 |
| E = 4, | z = 13 |

~~~~~/////~~~~~

*) Find the positive integers A, B, C, D, E, which must have a common factor of a prime number and the exponents u, v, x, y, z are greater than 2 of the following equation:

$$A^u + B^v + C^x + D^y = E^z$$

~~~~~Prove~~~~~

Apply the following formula, (a = 8 and n = 4)

$$\sum_{k \geq 2}^{n} a = a \cdot b^k$$

Substitute a = 8 and n = 4

$$\sum_{k \geq 2}^{n=4} 8 = 4 \cdot 8 = 32 = 2^5$$

Rewrite

$$\sum_{k\geq 2}^{n=4} 2^3 = 2^5$$

Multiply both sides by $2^{45}$

$$\sum_{k\geq 2}^{n=4} 2^3 = 2^5$$

$$2^{45}\sum_{k=2}^{n=4} 2^3 = 2^{45}\cdot 2^5$$

Writing simple form to the equation as follows

$$8 + 8 + 8 + 8 = 48 = 32 \qquad \text{(4c)}$$

Or

$$2^3 + 2^3 + 2^3 + 2^3 = 2^5$$

Multiply both sides of the equation (4c) by $2^{45}$

$$2^{48} + 2^{48} + 2^{48} + 2^{48} = 2^{50}$$

Rewriting the above equation in the form of Beal

$$(2^4)^{12} + (2^8)^6 + (2^2)^{24} + (2^6)^8 = (2^5)^{10}$$

Or

$$16^{12} + 256^{6} + 4^{24} + 64^{8} = 32^{10}$$

Solution

| | |
|---|---|
| A = 16, | u = 12 |
| B = 256, | v = 6 |
| C = 4, | x = 24 |
| D = 64, | y = 8 |
| E = 32, | z = 10 |

~~~~~/////~~~~~
~~~~~

# EXERCISE

*) Find the positive integers A, B, C, D, E, which must have a common factor of a prime number and the exponents u, v, x, y, z are greater than 2 of the following equation:

$$A^u + B^v + C^x + D^y = E^z$$

$u = 3, v = 5, x = 4$

*) Find the positive integers A, B, C, D, E, which must have a common factor of a prime number and the exponents u, v, x, y, z are greater than 2 of the following equation:

$$A^u + B^v + C^x + D^y = E^z$$

$u = 6, v = 5, x = 4$

*) Find the positive integers A, B, C, D, E, which must have a common factor of a prime number and the exponents u, v, x, y, z are greater than 2 of the following equation:

$$A^u + B^v + C^x + D^y = E^z$$

$u = 3, v = 5, x = 7$

*) Find the positive integers A, B, C, D, E, which must have a common factor of a prime number and the exponents u, v, x, y, z are greater than 2 of the following equation:

$$A^u + B^v + C^x + D^y = E^z$$

$u = 6, v = 7, x = 4$

*) Find the positive integers A, B, C, D, E, which must have a common factor of a prime number and the exponents u, v, x, y, z are greater than 2 of the following equation:

$$A^u + B^v + C^x + D^y = E^z$$

$u = 7, v = 5, x = 6$

*) Find the positive integers A, B, C, D, E, which must have a common factor of a prime number and the exponents u, v, x, y, z are greater than 2 of the following equation:

$$A^u + B^v + C^x + D^y = E^z$$

$u = 7, v = 5, x = 4$

*) Find the positive integers A, B, C, D, E, which must have a common factor of a prime number and the

exponents u, v, x, y, z are greater than 2 of the following equation:

$$A^u + B^v + C^x + D^y = E^z$$

$u = 3, v = 7, x = 4$

*) Find the positive integers A, B, C, D, E, which must have a common factor of a prime number and the exponents u, v, x, y, z are greater than 2 of the following equation:

$$A^u + B^v + C^x + D^y = E^z$$

$u = 3, v = 5, x = 7$

*) Find the positive integers A, B, C, D, E, which must have a common factor of a prime number and the exponents u, v, x, y, z are greater than 2 of the following equation:

$$A^u + B^v + C^x + D^y = E^z$$

$u = 6, v = 5, x = 7$

*) Find the positive integers A, B, C, D, E, which must have a common factor of a prime number and the exponents u, v, x, y, z are greater than 2 of the following equation:

$$A^u + B^v + C^x + D^y = E^z$$

$u = 7, v = 5, x = 5$

*) Find the positive integers A, B, C, D, E, which must have a common factor of a prime number and the exponents u, v, x, y, z are greater than 2 of the following equation:

$$A^u + B^v + C^x + D^y = E^z$$

$u = 6, v = 7, x = 4$

*) Find the positive integers A, B, C, D, E, which must have a common factor of a prime number and the exponents u, v, x, y, z are greater than 2 of the following equation:

$$A^u + B^v + C^x + D^y = E^z$$

$u = 3, v = 5, x = 4, y = 6$

*) Find the positive integers A, B, C, D, E, which must have a common factor of a prime number and the exponents u, v, x, y, z are greater than 2 of the following equation:

$$A^u + B^v + C^x + D^y = E^z$$

$u = 7, v = 5, x = 4, y = 6$

*) Find the positive integers A, B, C, D, E, which must have a common factor of a prime number and the exponents u, v, x, y, z are greater than 2 of the following equation:

$$A^u + B^v + C^x + D^y = E^z$$

$u = 7, v = 5, x = 6, y = 8$

*) Find the positive integers A, B, C, D, E, which must have a common factor of a prime number and the exponents u, v, x, y, z are greater than 2 of the following equation:

$$A^u + B^v + C^x + D^y = E^z$$

$u = 8$, $v = 5$, $x = 4$, $y = 7$

*) Find the positive integers A, B, C, D, E, which must have a common factor of a prime number and the exponents u, v, x, y, z are greater than 2 of the following equation:

$$A^u + B^v + C^x + D^y = E^z$$

$u = 3$, $v = 8$, $x = 7$, $y = 6$

# EXPANSION OF THE BEAL CONJECTURE TO FORM

$$A^t + B^u + C^v + D^x + E^y = F^z$$

We have completed proof of Beal Conjecture, so the hypothesis becomes Beal's Theorem. Now we use Beal's Theorem in the form of the following equation:

$$A^t + B^u + C^v + D^x + E^y = F^z$$

Values of A, B, C, D, E, F (positive integers), which must have a common factor of a prime number, and the exponents t, u, v, x, y, z are greater than 2

*) Find the positive integers A, B, C, D, E, F which must have a common factor of a prime number and the exponents t, u, v, x, y, z are greater than 2 of the following equation:

$$A^t + B^u + C^v + D^x + E^y = F^z$$

~~~~~Prove~~~~~

Apply the following formula (a = 1 and n = 5)

$$\sum_{k=1}^{n=5} a = b^1$$

Substitute a = 1

$$\sum_{k=1}^{n=5} 1 = 5^1$$

Multiply both sides by $b^m$

$$\sum_{k=1}^{n=5} 1 = 5^1$$

$$b^m \sum_{k=2}^{n=4} a = b^{m+1}$$

Given $b^m = 5^{48}$

$$5^{48} \sum_{k=1}^{n=5} 1 = 5^{49}$$
~~~~~

Rewrite

$$1 + 1 + 1 + 1 + 1 = 5 \qquad (1d)$$

Multiply both sides of the equation (1d) by $5^{48}$

$$5^{48} + 5^{48} + 5^{48} + 5^{48} + 5^{48} = 5^{49}$$

Rewriting the above equation in the form of Beal

$$(5^4)^{12} + (5^6)^8 + (5^2)^{24} + (5^3)^{16} + (5^8)^6 = (5^7)^7$$

$$625^{12} + 15625^8 + 25^{24} + 125^{16} + 390625^6 = 78125^7$$

Solution

$A = 625$, $t = 12$
$B = 15625$, $u = 8$
$C = 25$, $v = 24$
$D = 125$, $x = 16$
$E = 390625$, $y = 6$
$F = 78125$, $z = 7$

*) Find the positive integers A, B, C, D, E, F which must have a common factor of a prime number and the exponents t, u, v, x, y, z are greater than 2 of the following equation:

$$A^t + B^u + C^v + D^x + E^y = F^z$$

~~~~~Prove~~~~~

Apply the following formula (a = 5 and n = 5)

$$\sum_{k=2}^{n=5} a = a^2$$

a = 5 we have

$$\sum_{k=1}^{n=5} 5 = 5^2$$

Multiply both sides by $a^m$

$$a^m \sum_{k=2}^{n=5} a = a^{m+2}$$

Substitute $a^m = 5^{23}$

$$5^{23} \sum_{k=2}^{n=5} 5 = 5^{25}$$

Rewrite

$$5 + 5 + 5 + 5 + 5 = 25 = 5^2 \qquad (2d)$$

Multiply both sides of the equation (2d) by $5^{23}$

$5^{24} + 5^{24} + 5^{24} + 5^{24} + 5^{24} = 5^{25}$
~~~~~

$(5^4)^6 + (5^3)^8 + (5^2)^{12} + (5^6)^4 + (5^8)^3 = (5^5)^5$

Rewriting the above equation in the form of Beal

$625^6 + 125^8 + 25^{12} + 15625^4 + 390625^3 = 3125^5$

We have

| | |
|---|---|
| A = 625, | t = 6 |
| B = 125 | u = 8 |
| C = 25, | v = 12 |
| D = 15625, | x = 4 |
| E = 390625 | y = 3 |
| F = 3125 | z = 5 |

~~~~~~/////~~~~~~

*) Find the positive integers A, B, C, D, E, F which must have a common factor of a prime number and the exponents t, u, v, x, y, z are greater than 2 of the following equation:

$$A^t + B^u + C^v + D^x + E^y = F^z$$

~~~~~Prove~~~~~

Apply the following formula, (a = 1 and n = 5)

Similarly the equation (1d) above, but the solution is difference

$$\sum_{k=1}^{n=5} a = b$$

Substitute a = 1

$$\sum_{k=1}^{n=5} 1 = 5$$

Multiply both sides by $b^m$

$$b^m \sum_{k=1}^{n=5} a = b^{m+1}$$

Given $b^m = 5^{36}$

$$5^{36} \sum_{k=2}^{n=5} 1 = 5^{37}$$

Rewrite

$$1 + 1 + 1 + 1 + 1 = 5 \qquad (3d)$$

Multiply both sides of the equation (3d) by $5^{36}$

$$5^{36} + 5^{36} + 5^{36} + 5^{36} + 5^{36} = 5^{37}$$

Rewriting the above equation in the form of Beal

$$(5^3)^{12} + (5^4)^9 + (5^2)^{18} + (5^6)^6 + (5^{12})^3 = (5)^{37}$$

$$125^{12} + 625^{9} + 25^{18} + 15625^{6} + 244140625^{3} = 5^{37}$$

To make sure, we must try again

Left equation

$$125^{12} + 625^{9} + 25^{18} + 15625^{6} + 244140625^{3} = 7275957614183425903320312 5$$

Right equation

$$5^{37} = 72759576141834259033203125$$

Solution

| | |
|---|---|
| A = 125, | t = 12 |
| B = 625, | u = 9 |
| C = 25, | v = 18 |
| D = 15625, | x = 6 |
| E = 244140625, | y = 3 |
| F = 5, | z = 37 |

~~~~~/////~~~~~
~~~~~

# EXERCISE

*) Find the positive integers A, B, C, D, E, F, which must have a common factor of a prime number and the exponents t, u, v, x, y, z are greater than 2 of the following equation:

$$A^t + B^u + C^v + D^x + E^y = F^z$$

Given t = 3, u = 5, v = 4

*) Find the positive integers A, B, C, D, E, F, which must have a common factor of a prime number and the exponents t, u, v, x, y, z are greater than 2 of the following equation:

$$A^t + B^u + C^v + D^x + E^y = F^z$$

Given t = 7, u = 5, v = 4

*) Find the positive integers A, B, C, D, E, F, which must have a common factor of a prime number and the exponents t, u, v, x, y, z are greater than 2 of the following equation:

$$A^t + B^u + C^v + D^x + E^y = F^z$$

Given $t = 4$, $u = 6$, $v = 7$

*) Find the positive integers A, B, C, D, E, F which must have a common factor of a prime number and the exponents t, u, v, x, y, z are greater than 2 of the following equation:

$$A^t + B^u + C^v + D^x + E^y = F^z$$

Given $t = 6$, $u = 5$, $v = 5$

*) Find the positive integers A, B, C, D, E, F which must have a common factor of a prime number and the exponents t, u, v, x, y, z are greater than 2 of the following equation:

$$A^t + B^u + C^v + D^x + E^y = F^z$$

Given $t = 5$, $u = 4$, $v = 8$

*) Find the positive integers A, B, C, D, E, F, which must have a common factor of a prime number and the exponents t, u, v, x, y, z are greater than 2 of the following equation:

$$A^t + B^u + C^v + D^x + E^y = F^z$$

Given $t = 6$, $u = 5$, $v = 4$, $x = 3$

*) Find the positive integers A, B, C, D, E, F, which must have a common factor of a prime number and

the exponents t, u, v, x, y, z are greater than 2 of the following equation:

$$A^t + B^u + C^v + D^x + E^y = F^z$$

Given t = 8, u = 5, v = 4, x = 7

*) Find the positive integers A, B, C, D, E, F which must have a common factor of a prime number and the exponents t, u, v, x, y, z are greater than 2 of the following equation:

$$A^t + B^u + C^v + D^x + E^y = F^z$$

Given t = 7, u = 6, v = 4, x = 9

*) Find the positive integers A, B, C, D, E, F, which must have a common factor of a prime number and the exponents t, u, v, x, y, z are greater than 2 of the following equation:

$$A^t + B^u + C^v + D^x + E^y = F^z$$

Given t = 8, u = 9, v = 4, x = 3

*) Find the positive integers A, B, C, D, E, F, which must have a common factor of a prime number and the exponents t, u, v, x, y, z are greater than 2 of the following equation:

$$A^t + B^u + C^v + D^x + E^y = F^z$$

Given t = 10, u = 5, v = 4, x = 3

*) Find the positive integers A, B, C, D, E, F, which must have a common factor of a prime number and the exponents t, u, v, x, y, z are greater than 2 of the following equation:

$$A^t + B^u + C^v + D^x + E^y = F^z$$

Given t = 11, u = 5, v = 4, x = 8

*) Find the positive integers A, B, C, D, E, F, which must have a common factor of a prime number and the exponents t, u, v, x, y, z are greater than 2 of the following equation:

$$A^t + B^u + C^v + D^x + E^y = F^z$$

Given t = 10, u = 5, v = 7, x = 3

*) Find the positive integers A, B, C, D, E, F, which must have a common factor of a prime number and the exponents t, u, v, x, y, z are greater than 2 of the following equation:

$$A^t + B^u + C^v + D^x + E^y = F^z$$

Given t = 11, u = 5, v = 9, x = 3

*) Find the positive integers A, B, C, D, E, F, which must have a common factor of a prime number and the exponents t, u, v, x, y, z are greater than 2 of the following equation:

$$A^t + B^u + C^v + D^x + E^y = F^z$$

Given t = 6, u = 11, v = 7, x = 3

*) Find the positive integers A, B, C, D, E, F, which must have a common factor of a prime number and the exponents t, u, v, x, y, z are greater than 2 of the following equation:

$$A^t + B^u + C^v + D^x + E^y = F^z$$

Given t = 10, u = 9, v = 8, x = 3

*) Find the positive integers A, B, C, D, E, F, which must have a common factor of a prime number and the exponents t, u, v, x, y, z are greater than 2 of the following equation:

$$A^t + B^u + C^v + D^x + E^y = F^z$$

Given t = 12, u = 5, v = 6, x = 3

*) Find the positive integers A, B, C, D, E, F, which must have a common factor of a prime number and the exponents t, u, v, x, y, z are greater than 2 of the following equation:

$$A^t + B^u + C^v + D^x + E^y = F^z$$

Given t = 12, u = 5, v = 8, x = 7

*) Find the positive integers A, B, C, D, E, F, which must have a common factor of a prime number and

the exponents t, u, v, x, y, z are greater than 2 of the following equation:

$$A^t + B^u + C^v + D^x + E^y = F^z$$

Given t = 15, u = 7, v = 8, x = 4

*) Find the positive integers A, B, C, D, E, F, which must have a common factor of a prime number and the exponents t, u, v, x, y, z are greater than 2 of the following equation:

$$A^t + B^u + C^v + D^x + E^y = F^z$$

Given t = 13, u = 6, v = 9, x = 8

# EXPANSION OF THE BEAL CONJECTURE TO FORM

$$A^s + B^t + C^u + D^v + E^x + F^y = G^z$$

We have completed proof of Beal Conjecture, so the hypothesis becomes Beal's Theorem. Now we use Beal's Theorem in the form of the following equation:

$$A^s + B^t + C^u + D^v + E^x + F^y = G^z$$

Values of A, B, C, D, E, F, G (positive integers), which must have a common factor of a prime number, and the exponents s, t, u, v, x, y, z are greater than 2

*) Find the positive integers A, B, C, D, E, F, G, which must have a common factor of a prime number and the exponents s, t, u, v, x, y, z are greater than 2 of the following equation:

$$A^s + B^t + C^u + D^v + E^x + F^y = G^z$$

~~~~~Prove~~~~~

Similarly above we apply the following formula

a = 1 and n = 6

$$\sum_{a=1}^{n=6} a = b$$

Multiply both sides of the equation for $b^m$

$$b^m \sum_{k=1}^{n=6} a = b^{m+1}$$

Implementing the above formula, we have

$$1 + 1 + 1 + 1 + 1 + 1 = 6 \qquad (1e)$$

Multiply both sides of the equation (1e) for $6^{48}$

$$6^{48} + 6^{48} + 6^{48} + 6^{48} + 6^{48} + 6^{48} = 6^{49}$$
~~~~~

Rewrite to form of the Beal Conjecture

$(6)^{48} + (6^4)^{12} + (6^8)^6 + (6^{12})^4 + (6^2)^{24} + (6^3)^{16} = (6^7)^7$

Then

$6^{48} + 1296^{12} + 1679616^6 + 2176782336^4 + 36^{24} + 216^{16}$
$= 279936^7$

To make sure, we must try again

Left equation

$6^{48} + 1296^{12} + 1679616^6 + 2176782336^4 + 36^{24} + 216^{16}$
$= 1.3471354624412734344052326674276e+38$

Right equation

$279936^7 = 1.3471354624412734344052326674276e+38$

Solution

| | |
|---|---|
| A = 6, | s = 48 |
| B = 1296, | t = 12 |
| C = 1679616, | u = 6 |
| D = 2176782336, | v = 4 |
| E = 36 | x = 24 |
| F = 216, | y = 16 |
| G = 279936 | z = 7 |

*) Find the positive integers A, B, C, D, E, F, G, which must have a common factor of a prime number and the exponents s, t, u, v, x, y, z are greater than 2 of the following equation:

$$A^s + B^t + C^u + D^v + E^x + F^y = G^z$$

~~~~~Prove~~~~~

Similarly above we apply the following formula

$a = 6$ and $n = 6$

$$\sum_{k=2}^{n=6} a = a^2$$

Multiply both sides of the equation for $a^m$

$$a^m \sum_{k=2}^{n=6} a = a^{m+2}$$

Given $a^m = 6^{35}$

$$6^{35} \sum_{k=2}^{n=6} 6 = 6^{37}$$

Implementing the above formula, we have

$$6 + 6 + 6 + 6 + 6 + 6 = 6^2 \qquad \text{(2e)}$$
~~~~~

Multiply both sides of the equation (2e) for $6^{35}$

$$6^{36} + 6^{36} + 6^{36} + 6^{36} + 6^{36} + 6^{36} = 6^{37}$$

Rewrite to form of the Beal Conjecture

$$(6^4)^9 + (6^2)^{18} + (6^6)^6 + (6^9)^4 + (6^3)^{12} + (6^{12})^3 = (6)^{37}$$

Or

$$1296^9 + 36^{18} + 46656^6 + 10077696^4 + 216^{12} + 2176782336^3 = 6^{37}$$

Try again

Left

$$1296^9 + 36^{18} + 46656^6 + 10077696^4 + 216^{12} + 6^{36}$$
$$= 61886548790943213277031694336$$

Right

$$6^{37} = 61886548790943213277031694336$$

Solution

| | |
|---|---|
| A = 6, | s = 48 |
| B = 1296, | t = 12 |
| C = 1679616, | u = 6 |
| D = 2176782336, | v = 4 |
| E = 36, | x = 24 |
| F = 216, | y = 16 |
| G = 279936 | z = 7 |

~~~~~/////~~~~~
~~~~~

*) Find the positive integers A, B, C, D, E, F, G, which must have a common factor of a prime number and the exponents s, t, u, v, x, y, z are greater than 2 of the following equation:

$$A^s + B^t + C^u + D^v + E^x + F^y = G^z$$

~~~~~Prove~~~~~

Similarly above we apply the following formula

a = 2 and n = 6

$$\sum_{k=2}^{n=6} a = b \cdot a^2$$

Multiply both sides of the equation for $a^m \cdot b^q$

$$a^m \cdot b^q \sum_{k=2}^{n=6} a = a^{m+2} \cdot b^q$$

Given $a^m \cdot b^q = 2^{23} \cdot 3^{24}$

$$2^{23} \cdot 3^{24} \sum_{k=2}^{n=6} 2 = 2^{25} \cdot 3^{25}$$

Implementing the above formula, we have (3e)

$$2 + 2 + 2 + 2 + 2 + 2 = 12 = 2^2 \cdot 3 \qquad (3e)$$
~~~~~

Multiply both sides of the equation (3e) for $2^{23} \cdot 3^{24}$

$$2^{24} \cdot 3^{24} + 2^{24} \cdot 3^{24} + 2^{24} \cdot 3^{24} + 2^{24} \cdot 3^{24} + 2^{24} \cdot 3^{24} + 2^{24} \cdot 3^{24} = 2^{25} \cdot 3^{25}$$

Rewrite to form of the Beal Conjecture

$$6^{24} + 36^{12} + 216^{8} + 1296^{6} + 1679616^{3} + 46656^{4} = 6^{25}$$

Try again

Left

$$6^{24} + 36^{12} + 216^{8} + 1296^{6} + 1679616^{3} + 46656^{4} = 28430288029929701376$$

Right

$$6^{25} = 28430288029929701376$$

Solution

A = 6, s = 24
B = 36, t = 12
C = 216, u = 8
D = 1296, v = 6
E = 1679616, x = 3
F = 46656, y = 4
G = 6, z = 25

~~~~~/////~~~~~
~~~~~

# EXERCISE

*) Find the positive integers A, B, C, D, E, F, G, which must have a common factor of a prime number and the exponents s, t, u, v, x, y, z are greater than 2 of the following equation:

$$A^s + B^t + C^u + D^v + E^x + F^y = G^z$$

Given s = 3, t = 4, u = 5

*) Find the positive integers A, B, C, D, E, F, G, which must have a common factor of a prime number and the exponents s, t, u, v, x, y, z are greater than 2 of the following equation:

$$A^s + B^t + C^u + D^v + E^x + F^y = G^z$$

Given s = 3, t = 4, u = 5

*) Find the positive integers A, B, C, D, E, F, G, which must have a common factor of a prime number and the exponents s, t, u, v, x, y, z are greater than 2 of the following equation:

$$A^s + B^t + C^u + D^v + E^x + F^y = G^z$$

Given $s = 3, t = 4, u = 5$

*) Find the positive integers A, B, C, D, E, F, G, which must have a common factor of a prime number and the exponents s, t, u, v, x, y, z are greater than 2 of the following equation:

$$A^s + B^t + C^u + D^v + E^x + F^y = G^z$$

Given $s = 3, t = 4, u = 5$

*) Find the positive integers A, B, C, D, E, F, G, which must have a common factor of a prime number and the exponent s, t, u, v, x, y, z are greater than 2 of the following equation:

$$A^s + B^t + C^u + D^v + E^x + F^y = G^z$$

Given $s = 3, t = 4, u = 5$

*) Find the positive integers A, B, C, D, E, F, G, which must have a common factor of a prime number and the exponents s, t, u, v, x, y, z are greater than 2 of the following equation:

$$A^s + B^t + C^u + D^v + E^x + F^y = G^z$$

Given $s = 3, t = 4, u = 5$

*) Find the positive integers A, B, C, D, E, F, G, which must have a common factor of a prime number and

the exponents s, t, u, v, x, y, z are greater than 2 of the following equation:

$$A^s + B^t + C^u + D^v + E^x + F^y = G^z$$

Given s = 3, t = 4, u = 5

*) Find the positive integers A, B, C, D, E, F, G, which must have a common factor of a prime number and the exponents s, t, u, v, x, y, z are greater than 2 of the following equation:

$$A^s + B^t + C^u + D^v + E^x + F^y = G^z$$

Given s = 3, t = 4, u = 5

*) Find the positive integers A, B, C, D, E, F G, which must have a common factor of a prime number and the exponents s, t, u, v, x, y, z are greater than 2 of the following equation:

$$A^s + B^t + C^u + D^v + E^x + F^y = G^z$$

Given s = 3, t = 4, u = 5

*) Find the positive integers A, B, C, D, E, F, G, which must have a common factor of a prime number and the exponents s, t, u, v, x, y, z are greater than 2 of the following equation:

$$A^s + B^t + C^u + D^v + E^x + F^y = G^z$$

Given s = 3, t = 4, u = 5

*) Find the positive integers A, B, C, D, E, F, G, which must have a common factor of a prime number and the exponents s, t, u, v, x, y, z are greater than 2 of the following equation:

$$A^s + B^t + C^u + D^v + E^x + F^y = G^z$$

Given s = 3, t = 4, u = 5

*) Find the positive integers A, B, C, D, E, F, G, which must have a common factor of a prime number and the exponents s, t, u, v, x, y, z are greater than 2 of the following equation:

$$A^s + B^t + C^u + D^v + E^x + F^y = G^z$$

Given s = 3, t = 4, u = 5

*) Find the positive integers A, B, C, D, E, F, G, which must have a common factor of a prime number and the exponents s, t, u, v, x, y, z are greater than 2 of the following equation:

$$A^s + B^t + C^u + D^v + E^x + F^y = G^z$$

Given s = 3, t = 4, u = 5

*) Find the positive integers A, B, C, D, E, F, G, which must have a common factor of a prime number and the exponents s, t, u, v, x, y, z are greater than 2 of the following equation:

$$A^s + B^t + C^u + D^v + E^x + F^y = G^z$$

Given $s = 3$, $t = 4$, $u = 5$

*) Find the positive integers A, B, C, D, E, F, G, which must have a common factor of a prime number and the exponents s, t, u, v, x, y, z are greater than 2 of the following equation:

$$A^s + B^t + C^u + D^v + E^x + F^y = G^z$$

Given $s = 3$, $t = 4$, $u = 5$

*) Find the positive integers A, B, C, D, E, F, G, which must have a common factor of a prime number and the exponents s, t, u, v, x, y, z are greater than 2 of the following equation:

$$A^s + B^t + C^u + D^v + E^x + F^y = G^z$$

Given $s = 3$, $t = 4$, $u = 5$

*) Find the positive integers A, B, C, D, E, F, G, which must have a common factor of a prime number and the exponent s, t, u, v, x, y, z are greater than 2 of the following equation:

$$A^s + B^t + C^u + D^v + E^x + F^y = G^z$$

Given $s = 3$, $t = 4$, $u = 5$

*) Find the positive integers A, B, C, D, E, F, G, which must have a common factor of a prime number and

the exponents s, t, u, v, x, y, z are greater than 2 of the following equation:

$$A^s + B^t + C^u + D^v + E^x + F^y = G^z$$

Given s = 3, t = 4, u = 5

*) Find the positive integers A, B, C, D, E, F, G, which must have a common factor of a prime number and the exponents s, t, u, v, x, y, z are greater than 2 of the following equation:

$$A^s + B^t + C^u + D^v + E^x + F^y = G^z$$

Given s = 3, t = 4, u = 5

# EXPANSION OF THE BEAL CONJECTURE TO FORM

$$A^r + B^s + C^t + D^u + E^v + F^x + G^y = H^z$$

We have completed proof of Beal Conjecture, so the hypothesis becomes Beal's Theorem. Now we use the Beal's Theorem in the form of the following equation:

$$A^r + B^s + C^t + D^u + E^v + F^x + G^y = H^z$$

Values of A, B, C, D, E, F, G, H (positive integers), which must have a common factor of a prime number, and the exponents r, s, t, u, v, x, y, z are greater than 2

*) Find the positive integers A, B, C, D, E, F, G, H, which must have a common factor of a prime number and the exponents r, s, t, u, v, x, y, z are greater than 2 of the following equation:

$$A^r + B^s + C^t + D^u + E^v + F^x + G^y = H^z$$

~~~~~Prove~~~~~

Similarly above we apply the following formula

a = 7 and n = 7

$$\sum_{k=2}^{n=7} a = a^2$$

Given a = 7

$$\sum_{k=2}^{n=7} 7 = 7^2$$

Multiply 2 sides of the formula by $a^m$

$$a^m \sum_{k=2}^{n=7} a = a^{m+2}$$

Given $a^m = 7^{95}$

$$7^{95} \sum_{k=2}^{n=7} 7 = 7^{97}$$

Implementing the above formula, we have

$$7 + 7 + 7 + 7 + 7 + 7 + 7 = 49 = 7^2 \qquad (1f)$$
~~~~~

Multiply both sides of the equation (2f) for $7^{95}$

$$7^{96} + 7^{96} + 7^{96} + 7^{96} + 7^{96} + 7^{96} + 7^{96} = 7^{97}$$

Rewrite to form of the Beal Conjecture

$$(7^2)^{48} + (7^4)^{24} + (7^8)^{12} + (7^{12})^8 + (7^6)^{16} + (7^3)^{32} + (7^{16})^6 = (7)^{97}$$

Then

$$49^{48} + 2401^{24} + 5764801^{12} + 13841287201^8 + 117649^{16} + 343^{32} + 33232930569601^6 = 7^{97}$$

Try again

Left side

$$49^{48} + 2401^{24} + 5764801^{12} + 13841287201^8 + 117649^{16} + 343^{32} + 33232930569601^6 = 9.4299606694599358348240459740531e+81$$

Right side

$$7^{97} = 9.4299606694599358348240459740531e+81$$

Solution

| | |
|---|---|
| $A = 49$, | $r = 48$ |
| $B = 2401$, | $s = 24$ |
| $C = 5764801$, | $t = 12$ |
| $D = 13841287201$, | $u = 8$ |
| $E = 117649$, | $v = 16$ |

| | |
|---|---|
| $F = 343$, | $x = 32$ |
| $G = 33232930569601$, | $y = 6$ |
| $H = 7$ | $z = 97$ |

~~~~~/////~~~~~
~~~~~

# EXPANSION OF THE BEAL CONJECTURE TO FORM

# $A^q + B^r + C^s + D^t + E^u + F^v + G^x + H^y = I^z$

We have completed proof of the Beal Conjecture, so the hypothesis becomes Beal's Theorem. Now we use Beal's Theorem in the form of the following equation:

$$A^q + B^r + C^s + D^t + E^u + F^v + G^x + H^y = I^z$$

Values of A, B, C, D, E, F, G, H, I (positive integers), which must have a common factor of a prime number, and the exponents q, r, s, t, u, v, x, y, z are greater than 2

*) Find the positive integers A, B, C, D, E, F, G, H, I which must have a common factor of a prime

number and the exponents q, r, s, t, u, v, x, y, z are greater than 2 of the following equation:

$$A^q + B^r + C^s + D^t + E^u + F^v + G^x + H^y = I^z$$

~~~~~Prove~~~~~

Similarly above we apply the following formula

a = 2 và n = 8

$$\sum_{k=4}^{n=8} a = a^4$$

Put a = 2

$$\sum_{k=4}^{n=8} 2 = 16 = 2^4$$

Multiply both sides for 2 $a^m$

$$a^m \sum_{k=4}^{n=8} a = a^{m+4}$$

Given $a^m = 2^{71}$

$$2^{71} \sum_{k=4}^{n=8} 2 = 2^{75}$$
~~~~~

Implementing the above formula, we have

$$2 + 2 + 2 + 2 + 2 + 2 + 2 + 2 = 16 = 2^4 \qquad (1g)$$

Multiply both sides of the equation (1g) for $2^{71}$

$$2^{72} + 2^{72} + 2^{72} + 2^{72} + 2^{72} + 2^{72} + 2^{72} + 2^{72} = 2^{75} \qquad (1g)$$

Rewrite to form of the Beal Conjecture

$$(2^9)^8 + (2^2)^{36} + (2^4)^{18} + (2^8)^9 + (2^{12})^6 + (2^6)^{12} + (2^3)^{24} + (2^{18})^4 = (2^5)^{15}$$

Then

$$512^8 + 4^{36} + 16^{18} + 256^9 + 4096^6 + 64^{12} + 8^{24} + 262144^4 = 32^{15}$$

Try again

Left side

$$512^8 + 4^{36} + 16^{18} + 256^9 + 4096^6 + 64^{12} + 8^{24} + 262144^4$$
$$= 37778931862957161709568$$

Right side

$$32^{15} = 37778931862957161709568$$

Conclude

A = 512, q = 8
B = 4, r = 36
C = 16, s = 18
D = 256, t = 9

$E = 4096,$ $u = 6$
$F = 64,$ $v = 12$
$G = 8,$ $x = 24$
$H = 262144,$ $y = 4$
$I = 32,$ $z = 15$

*) Find the positive integers A, B, C, D, E, F, G, H, I which must have a common factor of a prime number and the exponents q, r, s, t, u, v, x, y, z are greater than 2 of the following equation:

$$A^q + B^r + C^s + D^t + E^u + F^v + G^x + H^y = I^z$$

~~~~~Prove~~~~~

Similarly above we apply the following formula

a = 3 and n = 8

$$\sum_{k=3}^{n=8} a = a \cdot b^3$$

Put a = 3 we have

$$\sum_{k=3}^{n=8} 3 = 3 \cdot 8 = 3 \cdot 2^3$$
~~~~~

Multiply both sides for $a^m$ $b^q$

$$a^m b^q \sum_{k=3}^{n=8} a = a^{m+1} b^{q+3}$$

Given $a^m$ $b^q = 3^{83}$ $2^{84}$

$$3^{83} 2^{84} \sum_{k=3}^{n=8} 3 = 3^{84} 2^{87}$$

Implementing the above formula, we have

$3 + 3 + 3 + 3 + 3 + 3 + 3 + 3 = 24 = 3 \cdot 8 = 3 \cdot 2^3$ (2g)

Multiply both sides of the equation (2g) for $3^{83} \cdot 2^{84}$

$$3^{84} \cdot 2^{84} + 3^{84} \cdot 2^{84} + 3^{84} \cdot 2^{84} + 3^{84} \cdot 2^{84} + 3^{84} \cdot 2^{84} + 3^{84} \cdot 2^{84} + 3^{84} \cdot 2^{84} + 3^{84} \cdot 2^{84} = 3^{84} \cdot 2^{87}$$

Rewrite

$$(6)^{84} + (3^{21} \cdot 2^{21})^4 + (3^4 \cdot 2^4)^{21} + (3^{14} \cdot 2^{14})^6 + (3^7 \cdot 2^7)^{12} + (3^6 \cdot 2^6)^{14} + (3^{12} \cdot 2^{12})^7 + (3^3 \cdot 2^3)^{28} = (3^{28} \cdot 2^{29})^3$$

Then

$$6^{84} + 21936950640377856^4 + 1296^{21} + 78364164096^6 + 279936^{12} + 46656^{14} + 2176782336^7 + 216^{28} = 122818844289296309944 32^3$$

Try again

Left side

$6^{84} + 21936950640377856^{4} + 1296^{21} + 78364164096^{6} + 279936^{12} + 46656^{14} + 2176782336^{7} + 216^{28} =$
$= 1.8526569894307048213818919245369e+66$

Right side

$1228188442892963099443 2^{3}$
$= 1.8526569894307048213818919245369e+66$

Conclude

| | |
|---|---|
| A = 6, | q = 84 |
| B = 21936950640377856, | r = 4 |
| C = 1296, | s = 21 |
| D = 78364164096, | t = 6 |
| E = 279936, | u = 12 |
| F = 46656, | v = 14 |
| G = 2176782336, | x = 7 |
| H = 216, | y = 28 |
| I = 12281884428929630994432, | z = 3 |

~~~~~/////~~~~~
~~~~~

# EXERCISE

*) Find the positive integers A, B, C, D, E, F, G, H, I which must have a common factor of a prime number and the exponents q, r, s, t, u, v, x, y, z are greater than 2 of the following equation:

$$A^q + B^r + C^s + D^t + E^u + F^v + G^x + H^y = I^z$$

Given GCD(q,r,s) = 1

*) Find the positive integers A, B, C, D, E, F, G, H, I which must have a common factor of a prime number and the exponents q, r, s, t, u, v, x, y, z are greater than 2 of the following equation:

$$A^q + B^r + C^s + D^t + E^u + F^v + G^x + H^y = I^z$$

Given GCD(q,r,s) = 1

*) Find the positive integers A, B, C, D, E, F, G, H, I which must have a common factor of a prime number and the exponents q, r, s, t, u, v, x, y, z are greater than 2 of the following equation:

$$A^q + B^r + C^s + D^t + E^u + F^v + G^x + H^y = I^z$$

Given GCD(q,r,s,t) = 1

*) Find the positive integers A, B, C, D, E, F, G, H, I which must have a common factor of a prime number and the exponents q, r, s, t, u, v, x, y, z are greater than 2 of the following equation:

$$A^q + B^r + C^s + D^t + E^u + F^v + G^x + H^y = I^z$$

Given GCD(q,r,s,t) = 1

*) Find the positive integers A, B, C, D, E, F, G, H, I which must have a common factor of a prime number and the exponents q, r, s, t, u, v, x, y, z are greater than 2 of the following equation:

$$A^q + B^r + C^s + D^t + E^u + F^v + G^x + H^y = I^z$$

Given GCD(q,r,s,t,u) = 1

*) Find the positive integers A, B, C, D, E, F, G, H, I which must have a common factor of a prime number and the exponents q, r, s, t, u, v, x, y, z are greater than 2 of the following equation:

$$A^q + B^r + C^s + D^t + E^u + F^v + G^x + H^y = I^z$$

Given GCD(x,y,z) = 1

*) Find the positive integers A, B, C, D, E, F, G, H, I which must have a common factor of a prime

number and the exponents q, r, s, t, u, v, x, y, z are greater than 2 of the following equation:

$$A^q + B^r + C^s + D^t + E^u + F^v + G^x + H^y = I^z$$

Given GCD(x,y,z) = 1

*) Find the positive integers A, B, C, D, E, F, G, H, I which must have a common factor of a prime number and the exponents q, r, s, t, u, v, x, y, z are greater than 2 of the following equation:

$$A^q + B^r + C^s + D^t + E^u + F^v + G^x + H^y = I^z$$

Given GCD(v,x,y,z) = 1

*) Find the positive integers A, B, C, D, E, F, G, H, I which must have a common factor of a prime number and the exponents q, r, s, t, u, v, x, y, z are greater than 2 of the following equation:

$$A^q + B^r + C^s + D^t + E^u + F^v + G^x + H^y = I^z$$

Given GCD(v,x,y,z) = 1

# EXPANSION OF THE BEAL CONJECTURE TO FORM

$$A^p + B^q + C^r + D^s + E^t + F^u + G^v + H^x + I^y = J^z$$

We have completed proof of the Beal Conjecture, so the hypothesis becomes Beal's Theorem. Now we use Beal's Theorem in the form of the following equation

$$A^p + B^q + C^r + D^s + E^t + F^u + G^v + H^x + I^y = J^z$$

Values of A, B, C, D, E, F, G, H, I, J (positive integers), which must have a common factor of a prime number, and the exponents p, q, r, s, t, u, v, x, y, z are greater than 2

*) Find the positive integers A, B, C, D, E, F, G, H, I, J which must have a common factor of a prime

number and the exponents p, q, r, s, t, u, v, x, y, z are greater than 2 of the following equation:

$$A^p + B^q + C^r + D^s + E^t + F^u + G^v + H^x + I^y = J^z$$

~~~~~Prove~~~~~

We apply the following formula

a = 3 and n = 9

$$\sum_{k\geq 2}^{n} a = a^m$$

Put n= 9 and a= 3
We have

$$\sum_{k\geq 2}^{n=9} 3 = 9\cdot 3 = 27 = 3^3$$

Multiply both sides for $3^{83}$

$$3^{83}\sum_{k\geq 2}^{n} 3 = 3^{86}$$

Implementing the above formula, we have

$$3 + 3 + 3 + 3 + 3 + 3 + 3 + 3 + 3 = 27 = 3^3 \qquad \text{(1h)}$$
~~~~~

Multiply both sides of the equation (1h) for $3^{83}$

$$3^{84} + 3^{84} + 3^{84} + 3^{84} + 3^{84} + 3^{84} + 3^{84} + 3^{84} + 3^{84} = 3^{86}$$

Rewrite to form of the Beal Conjecture

$$27^{28} + 81^{21} + 531441^{7} + 4782969^{6} + 10460353203^{4} + 22876792454961^{3} + 729^{14} + 2187^{12} + 9^{42} = 9^{43}$$

$$(2^9)^8 + (2^2)^{36} + (2^4)^{18} + (2^8)^9 + (2^{12})^6 + (2^6)^{12} + (2^3)^{24} + (2^{18})^4 = (2^5)^{15}$$

Then

$$512^8 + 4^{36} + 16^{18} + 256^9 + 4096^6 + 64^{12} + 8^{24} + 262144^4 = 32^{15}$$

Try again

Left side

$$512^8 + 4^{36} + 16^{18} + 256^9 + 4096^6 + 64^{12} + 8^{24} + 262144^4 = 37778931862957161709568$$

Right side

$$32^{15} = 37778931862957161709568$$

Conclude

$A = 512$, $q = 8$
$B = 4$, $r = 36$
$C = 16$, $s = 18$
$D = 256$, $t = 9$
$E = 4096$, $u = 6$

F = 64,
G = 8,
H = 262144,
I = 32

v = 12
x = 24
y = 4
z = 15

~~~~~/////~~~~~
~~~~~

# EXERCISE

*) Find the positive integers A, B, C, D, E, F, G, H, I, J which must have a common factor of a prime number and the exponents p, q, r, s, t, u, v, x, y, z are greater than 2 of the following equation:

$$A^p + B^q + C^r + D^s + E^t + F^u + G^v + H^x + I^y = J^z$$

Given gcd(p,q,r) = 1

*) Find the positive integers A, B, C, D, E, F, G, H, I, J which must have a common factor of a prime number and the exponents p, q, r, s, t, u, v, x, y, z are greater than 2 of the following equation:

$$A^p + B^q + C^r + D^s + E^t + F^u + G^v + H^x + I^y = J^z$$

Given gcd(p,q,r) = 1

*) Find the positive integers A, B, C, D, E, F, G, H, I, J which must have a common factor of a prime number and the exponents p, q, r, s, t, u, v, x, y, z are greater than 2 of the following equation:

$$A^p + B^q + C^r + D^s + E^t + F^u + G^v + H^x + I^y = J^z$$

Given gcd(p,q,r) = 1

*) Find the positive integers A, B, C, D, E, F, G, H, I, J which must have a common factor of a prime number and the exponents p, q, r, s, t, u, v, x, y, z are greater than 2 of the following equation:

$$A^p + B^q + C^r + D^s + E^t + F^u + G^v + H^x + I^y = J^z$$

Given gcd(p,q,r,s) = 1

*) Find the positive integers A, B, C, D, E, F, G, H, I, J which must have a common factor of a prime number and the exponents p, q, r, s, t, u, v, x, y, z are greater than 2 of the following equation:

$$A^p + B^q + C^r + D^s + E^t + F^u + G^v + H^x + I^y = J^z$$

Given gcd(p,q,r,s) = 1

*) Find the positive integers A, B, C, D, E, F, G, H, I, J which must have a common factor of a prime number and the exponents p, q, r, s, t, u, v, x, y, z are greater than 2 of the following equation:

$$A^p + B^q + C^r + D^s + E^t + F^u + G^v + H^x + I^y = J^z$$

Given gcd(p,q,r,s) = 1

*) Find the positive integers A, B, C, D, E, F, G, H, I, J which must have a common factor of a prime number

number and the exponents p, q, r, s, t, u, v, x, y, z are greater than 2 of the following equation:

$$A^p + B^q + C^r + D^s + E^t + F^u + G^v + H^x + I^y = J^z$$

Given gcd(x,y,z) = 1

*) Find the positive integers A, B, C, D, E, F, G, H, I, J which must have a common factor of a prime number and the exponents p, q, r, s, t, u, v, x, y, z are greater than 2 of the following equation:

$$A^p + B^q + C^r + D^s + E^t + F^u + G^v + H^x + I^y = J^z$$

Given gcd(x,y,z) = 1

*) Find the positive integers A, B, C, D, E, F, G, H, I, J which must have a common factor of a prime number and the exponents p, q, r, s, t, u, v, x, y, z are greater than 2 of the following equation:

$$A^p + B^q + C^r + D^s + E^t + F^u + G^v + H^x + I^y = J^z$$

Given gcd(v,x,y,z) = 1

# EXPANSION OF THE BEAL CONJECTURE TO FORM

$$A^m + B^n + C^o + D^p + E^q + F^r + G^s + H^t + I^u + J^v + K^x + L^y = M^z$$

We have completed proof of the Beal Conjecture, so the hypothesis becomes Beal's Theorem. Now we use Beal's Theorem in the form of the following equation

$$A^m + B^n + C^o + D^p + E^q + F^r + G^s + H^t + I^u + J^v + K^x + L^y = M^z$$

Values of A, B, C, D, E, F, G, H, I, J, K, L, M (positive integers), which must have a common factor of a prime number, and the exponents m, n, o, p, q, r, s, t, u, v, x, y, z are greater than 2

*) Find the positive integers A, B, C, D, E, F, G, H, I, J, K, L, M, which must have a common factor of a prime number and the exponents m, n, o, p, q, r, s, t, u, v, x, y, z are greater than 2 of the following equation:

$$A^m + B^n + C^o + D^p + E^q + F^r + G^s + H^t + I^u + J^v + K^x + L^y = M^z$$

~~~~~Prove~~~~~

We apply the following formula

a = 12 and n = 12

$$\sum_{k=2}^{n} a = a^2$$

Put a = 12 and n = 12

We have

$$\sum_{k=2}^{n=12} 12 = 12^2$$

Implementing the above formula, we have

$$12 + 12 + 12 + 12 + 12 + 12 + 12 + 12 + 12 + 12 + 12 + 12 = 12^2 \qquad (1i)$$
~~~~~

Multiply both sides for

$$a^m \sum_{k=2}^{n=12} a = a^{m+2}$$

Given $a^m = 12^{419}$

$$12^{419} \sum_{k=2}^{n=12} 12 = 12^{421}$$

Multiply both sides of the equation (1i) for $12^{419}$

$$12^{420} + 12^{420} + 12^{420} + 12^{420} + 12^{420} + 12^{420} + 12^{420} + 12^{420} + 12^{420} + 12^{420} + 12^{420} + 12^{420} = 12^{421}$$

Rewrite to form of the Beal Conjecture

$$(12^4)^{105} + (12^3)^{140} + (12^6)^{70} + (12^7)^{60} + (12^{10})^{42} + (12^{14})^{30} + (12^{15})^{28} + (12^{20})^{21} + (12^{21})^{20} + (12^5)^{84} + (12^2)^{210} + (12^{28})^{15} = 12^{421}$$

Then

$$20736^{105} \quad 1728^{140} + 2985984^{70} + 35831808^{60} + 61917364224^{42} + 144^{210} + 1283918464548864^{30} + 164844\,6623609512543951043690496^{15} + 15407021574586368^{28} + 3833759992447475122176^{21} + 248832^{84}\,46005119909369701466112^{20} = 12^{421}$$

Try again

Left side

$20736^{105}$ $1728^{140}$ + $2985984^{70}$ + $35831808^{60}$ + $61917364224^{42}$ + $144^{210}$ + $1283918464548864^{30}$ + $1648446623609512543951043690496^{15}$ + $15407021574586368^{28}$ + $3833759992447475122176^{21}$ + $248832^{84}$ $46005119909369701466112^{20}$ =
2.1642358465820252830662510119878e+454

Right side

$12^{421}$ = 2.1642358465820252830662510119878e+454

Conclude

| | |
|---|---|
| A = 20736, | m = 105 |
| B = 1728, | n = 140 |
| C = 2985984, | o = 70 |
| D = 35831808, | p = 60 |
| E = 61917364224, | q = 42 |
| F = 144, | r = 210 |
| G = 1283918464548864, | s = 30 |
| H = 1648446623609512543951043690496, | t = 15 |
| I = 15407021574586368, | u = 28 |
| J = 3833759992447475122176, | v = 21 |
| K = 248832, | x = 84 |
| L = 46005119909369701466112, | y = 20 |
| M = 12, | z = 421 |

# EXERCISE

*) Find the positive integers A, B, C, D, E, F, G, H, I, J, K, L, M, which must have a common factor of a prime number and the exponents m, n, o, p, q, r, s, t, u, v, x, y, z are greater than 2 of the following equation:

$$A^m + B^n + C^o + D^p + E^q + F^r + G^s + H^t + I^u + J^v + K^x + L^y = M^z$$

Given (m, n, o, p, q, r, s, t, u, v, x, y, z )

$$\{gcf(A,B,C,D,E,F,G,H,I,J,K,L,M) = P\}$$

*) Find the positive integers A, B, C, D, E, F, G, H, I, J, K, L, M, which must have a common factor of a prime number and the exponents m, n, o, p, q, r, s, t, u, v, x, y, z are greater than 2 of the following equation:

$$A^m + B^n + C^o + D^p + E^q + F^r + G^s + H^t + I^u + J^v + K^x + L^y = M^z$$

Given (m, n, o, p, q, r, s, t, u, v, x, y, z )

$$\{gcf(A,B,C,D,E,F,G,H,I,J,K,L,M) = P\}$$

*) Find the positive integers A, B, C, D, E, F, G, H, I, J, K, L, M, which must have a common factor of a prime number and the exponents m, n, o, p, q, r, s, t, u, v, x, y, z are greater than 2 of the following equation:

$$A^m + B^n + C^o + D^p + E^q + F^r + G^s + H^t + I^u + J^v + K^x + L^y = M^z$$

Given (m, n, o, p, q, r, s, t, u, v, x, y, z )

$$\{gcf(A,B,C,D,E,F,G,H,I,J,K,L,M) = P\}$$

*) Find the positive integers A, B, C, D, E, F, G, H, I, J, K, L, M, which must have a common factor of a prime number and the exponents m, n, o, p, q, r, s, t, u, v, x, y, z are greater than 2 of the following equation:

$$A^m + B^n + C^o + D^p + E^q + F^r + G^s + H^t + I^u + J^v + K^x + L^y = M^z$$

Given (m, n, o, p, q, r, s, t, u, v, x, y, z )

$$\{gcf(A,B,C,D,E,F,G,H,I,J,K,L,M) = P\}$$